William Hewson

Experimental Inquiries

: part the first : containing an inquiry into the properties of the blood : with remarks on some of its morbid appearances : and an appendix, relating to the discovery of the lymphatic system in birds, fish, and the animals called Amphibious

William Hewson

Experimental Inquiries

: part the first : containing an inquiry into the properties of the blood : with remarks on some of its morbid appearances : and an appendix, relating to the discovery of the lymphatic system in birds, fish, and the animals called Amphibious

ISBN/EAN: 9783337393151

Printed in Europe, USA, Canada, Australia, Japan

Cover: Foto ©berggeist007 / pixelio.de

More available books at **www.hansebooks.com**

Experimental Inquiries:

PART THE FIRST.

CONTAINING AN

INQUIRY

INTO THE

PROPERTIES OF THE BLOOD.

WITH

REMARKS on some of its MORBID APPEARANCES:

AND

An APPENDIX,

RELATING TO

The Discovery of the Lymphatic System in Birds, Fish, and the Animals called Amphibious.

The SECOND EDITION.

By WILLIAM HEWSON, F.R.S.
AND TEACHER OF ANATOMY.

Vere scire, est per causas scire.   Lord BACON.

LONDON:
Printed for J. JOHNSON, No. 72, St. Paul's Church-Yard.
MDCCLXXIV.

TO

Sir JOHN PRINGLE, Bart.

PHYSICIAN TO HER MAJESTY.

SIR,

I HAVE taken the liberty to addreſs you upon this occaſion, not ſo much that I might pay a compliment to your merit, as that I might boaſt of your friendſhip: It is impoſſible that I ſhould add any thing to your reputation by diſplaying your abilities; but you may add greatly to mine, even by detecting my miſtakes. You have done me the honour to approve this little Eſſay, and I am ambitious to tell the world, that it owes much

much both to your knowledge and your kindnefs; which have concurred to fuggeft many ufeful hints and judicious corrections, that make it more worthy of attention, and afford me a public opportunity of affuring you that I am, with the trueft refpect and efteem,

SIR,

Your moft obedient,

and moft obliged

humble fervant,

WILLIAM HEWSON.

PREFACE.

THE knowledge of the human frame, the prefervation of health, and the cure of difeafes, are objects of too great importance to mankind, for the Author of thefe fheets to doubt, that any attempts to promote them, how fmall foever, fhould not meet with a candid and indulgent reception from the Public. An Inquiry into the Properties of the Blood, it is prefumed, will be thought, in a particular manner, interefting, fince there is no part of the human body upon which more phyfiological reafoning is founded, nor any from which more inferences are drawn for the cure of difeafes. And, as the inquiry is made by experiments upon the blood as near as poffible to the ftate in which it circulates in the veffels, it is hoped that the conclusions made from them will stand the test of a candid examination, and lead to further observations and improvements.

Since the publication of the first Edition, some new experiments have been made, and a new chapter has been added, which contains a recapitulation of the principal facts and conclusions that are met with in this Essay. These additions are between p. 98 and 140.

The Appendix is a vindication of the Author's right to the discovery of the Lymphatic Vessels, in opposition to the claim of the learned Dr. Alexander Monro, Professor of Anatomy in the University of Edinburgh.

CONTENTS.

THE blood, on being taken from the veins, firſt coagulates, p. 1.—then ſeparates into craſſamentum and ſerum, 2.—The coagulation takes place even in the animal heat, 3, 4.—and ſo does the ſeparation, 5. The craſſamentum conſiſts of the coagulable lymph and red globules, *ibid.*——The coagulable lymph and ſerum, how differing, 5, 6.—The ſurface of the craſſamentum becoming florid, how explained, 7, 8.—Arterial blood, its colour different from that of the venous, 9.—This difference where taking place, and where loſt, 10.—Effects of neutral ſalts on the colour of the blood, *ibid.*—Their effects in preventing its coagulation, and how explained, 11, 12, 13, 14.—Common ſalt,

why

why used in such large quantities in preparing blood for culinary purposes, 14.—Different morbid appearances of the coagulable lymph, 15.—The coagulation of the lymph out of the body, to what owing, 16, 17.—Not owing to rest, 18.—nor to cold, 19.—but to air, 20.—Coagulates slowly by rest, in the veins, 21.—Experiments shewing this, 22, 23, 24.—It coagulates at different periods in different constitutions, and in different diseases, 25.—The lymph, how filling the sacs of aneurysms, 26.—How filling the extremities of arteries after amputation, 27.—and how forming moles or false conceptions, *ibid*.—Blood frozen, and thawed without being coagulated, 28.—Coagulable lymph, by what degree of heat fixed, 29, 30, 31.—The serum, by what degree coagulated, 32.—The inflammatory crust or size not formed of the serum, but of the coagulable lymph, 34, to 38.—Inflammation does not increase the disposition of the blood to coagulate, but lessens it, 39, to 43.

—The

—The blood, how coagulating in the hearts of dead animals, 44.—Inflammation does not thicken the blood, but thins it, 45, 46.—The serum not sensibly attenuated by inflammation, 47.—Specific gravity of the red globules not sensibly increased, 48.—The coagulable lymph is so much attenuated by inflammation as to dilute the serum, 49.—The inflammatory crust or size, how formed, 50.—It is not a certain sign of inflammation, 51.—The size appearing in the first cup, and not in the last, 52, to 56.—This explained, by shewing that the properties of the lymph are changed, even in the time of bleeding, 56, 57.—The time at which the blood coagulates in the first cup on being compared with that in the last, a criterion of the change produced on the body by blood-letting, 58; and again, 67, 68.—The partial size, how explained, 59.—The blood of people in health, when coagulating on exposition to air, 40; and again, 60.—

Its

Its difpofition to coagulate, increafed by weakening the body, 60, 61, 62.—Hemorrhages, how ſtopt, 63, 64.—The faintneſs attending them not to be counteracted, *ibid.*—Sudden evacuations, how contributing to ſtop hemorrhages, 65, 66; and again, 131.—The ſize fufpected to arife, in fome cafes, merely from a temporary exertion of ſtrength, 66, 67.—The craffamentum forming a bag, how explained, 69, 70.—Inſtance of the blood's coagulating very flowly, 71.—Cold, its effects in leffening the difpofition of the blood to coagulate, and how proved, 72, to 76 —It entirely prevents coagulation, 76.—The blood in cold animals that fleep during the winter does not coagulate, and why, 77.—An inſtance of ſudden changes produced on the coagulation of the blood, 80, 81, 82.—Additional proofs that the difpofition of the lymph to coagulate is leffened where the ſize appears, 83.—Alfo, that the ſize is occafioned by a ſtrong action of the veffels,

CONTENTS. xiii

veſſels, 84.—and is therefore removed by weakening them, *ibid.*—The appearance of the ſize in the firſt and laſt cups, but not in the ſecond or third, how explained, 85.—The ſize appearing, or not, according to the ſtrength with which the veſſels act, 86.—The ſize, ſingly conſidered, not a ſure indication of the neceſſity of blood-letting, 87.—Faintneſs and languor, their ſudden effects in thickening the blood, and in leſſening its diſpoſition to coagulate, 87, 88.—They ſhould therefore be promoted in hemorrhages, and by what means, 89.—The lymph ſometimes has its diſpoſition to coagulate leſſened, without being thinned, 90, 91, 92.—The ſize differs in denſity in different caſes, and how explained, 93.—The bottom of the craſſamentum why ſo ſoft when the blood is very ſizy, 94.—Bleeding, in the ordinary quantity, does not always weaken the body, nor change the properties of the blood, 95, 96.—A ſmall orifice improper, where weakneſs

is

is to be fuddenly produced by bleeding, 96, 97.—The blood that trickles down the arm is fometimes without fize, even in inflammatory complaints, and why, 97, 98.—The want of fize on the blood that has trickled down the arm not owing to its having been more expofed to the air, and more cooled, 98, to 105.—The blood that drops upon the plate why later in coagulating than that in the cups which ftand near it, 106.—The different appearances of the blood in the different cups not owing to any alteration in the orifice of the vein, 109.—A ligature upon the arm, when long continued, may produce a fmall quantity of fize, 110, to 116.—Inftances of the blood's not coagulating when expofed to the air, 117.—The blood in tumors fometimes does not coagulate, and why, 118, 119.——Recapitulation of the principal facts mentioned·in this Effay, 120, to 133.—The fize not the caufe but the effect of inflammation, 125.—The ferum, its properties, 134.—It

confifts

confifts of, what, 135.—How differing from the coagulable lymph, *ibid.*—Serofity, what 136.—The mucilage of the ferum may be infpiffated and again diffolved without coagulating, 137.—in which circumftance it agrees with the white of an egg, but differs from the coagulable lymph, 138.—Serum by being diluted with water, is remarkably altered and made to approach to the nature of milk, *ibid.*—Serum, as well as milk, coagulated by rennet, 139.—The ferum of the blood is not always tranfparent, but fometimes of the colour of whey, fometimes has a cream on its furface, and fometimes is as white as milk, 141.—In thefe laft cafes only it contains globules, and of what fort, *ibid.*—Inftances of milk-like ferum from authors, 142.—Cafes lately communicated, 143, to 147.—This appearance is not owing to unaffimilated chyle, *ibid.*—but to the fat's being re-abforbed, 149.—The re-abforption of fat, and its accumulation in in the blood-veſſels, ſuſpected to be a cauſe of plethora, 150.—The chyle in birds not white, *ibid.*—Fat, a new ſubſtance formed in the cellular membrane, and not a mere depoſition of the oily part of the food, 151.—The ſuperfluous food, why converted into fat, *ibid.*—Suet the moſt nutritive of all ſubſtances, *ibid.*—The whiteneſs of the ſerum owing to an extraordinary re-abſorption of the fat, which is a cauſe of want of appetite, &c. and not the effect, 152, 153, 154.—The whiteneſs of the ſerum to be attended to in ſome complaints, and why, 155.— APPENDIX, 159.—

Propoſals for a Courſe of Anatomical and Chirurgical Lectures, p. 219.

A N

AN

EXPERIMENTAL INQUIRY

INTO THE

PROPERTIES of the BLOOD.

CHAP. I.

Of the separation of the Serum; *the colour of the* Crassamentum; *and of the causes of the* COAGULATION OF THE BLOOD.

WHEN fresh blood is received into a bason, and suffered to rest, in a few minutes it jellies, or coagulates, and soon after separates into two parts, distinguished by the names of *Crassamentum* and *Serum.* These two parts differ in their proportions in different conftitutions; in a ftrong perfon, the *craſſamentum* is in greater proportion to the *ſerum* than in a weak one; and the fame difference is found to take place in difeafes; thence is deduced the general conclufion, that the lefs the quantity of *ſerum* is in proportion to the *craſſamentum*, bleeding, diluting liquors, and a low diet, are the more neceſſary; whilft in fome dropfies, and other difeafes where the *ſerum* is in a great, and the *craſſamentum* in a fmall proportion, bleeding and diluting would be highly improper. As it is therefore fuppofed ufeful to attend to the proportions of thefe parts in many diforders, and even to take indications of cure from them, it has been an object with thofe who have made experiments on the blood, to determine the circumftances on which its more perfect feparation into thefe two parts depends; it being obvious, that till this be done, our inferences from their proportions will be liable to confiderable fallacies. Two of the

the lateſt writers on this ſubject agree, that if the blood, after being taken from a vein, be ſet in a cold place, it will not eaſily ſeparate, and that a moderate warmth is neceſſary: this is a fact that is evinced by daily experience. They likewiſe ſay, that the heat ſhould be leſs than that of the animal, or than 98° of Fahrenheit's thermometer; and that, if freſh blood be received into a cup, and that cup put into water heated to 98°, it will not ſeparate; nay, they even ſay, that it will not coagulate; but this, I am perſuaded from experiments, is ill founded.

EXPERIMENT I.

A TIN-VESSEL, containing water, was placed upon a lamp, which kept the water in a heat that varied between 100 and 105 degrees. In this water was placed a phial, containing blood that inſtant taken from the arm of a perſon in health; the phial was previously warmed, then filled, and corked to exclude air. In the same water was placed a tea-cup half full of blood, just taken from the same person; a third portion of the blood was then received from the same vein into a bason, and was set upon a table, the heat of the atmosphere being at 67°. Now, according to their opinion, the two former should neither have coagulated nor separated, when that in the bason began to separate; but, on the contrary, they were all three found to coagulate nearly in the same time; and those in the warm water not only did separate as well as the other, but even sooner.

EXPERIMENT II.

The same experiment was repeated on the blood of a person that laboured under the acute rheumatism, whilst the heat of the atmosphere was no higher than 55°, and that of the warm water was 108°; and

and the result of this experiment was not only a confirmation of what was observed in the first, but it even shewed, that this degree of heat was so far from lessening, that it increased the disposition to coagulate; for the blood in the cup and in the phial was not only congealed, but the separation was much advanced before the whole of the blood in the bason was coagulated. Thence I am led to conclude, that the separation of the blood in a given time, is in proportion as the heat in which it stands is nearer to the animal heat, or 98°; or greater in that heat than in any of a less degree. And I am confirmed in this inference by experiments hereafter to be related, where the blood in the living animal, whilst at rest, was found both to coagulate and to separate.

It is well known, that the *crassamentum* consists of two parts, of which one gives it solidity, and is by some called the fibrous part of the blood, or the *gluten*, but by others with more propriety termed the *coagulable lymph*; and of another, which gives the red colour to the blood, and is called the *red globules*. These two parts can be separated by washing the *crassamentum* in water, the red particles dissolving in the water, whilst the coagulable lymph remains solid. That it is the coagulable lymph, which, by its becoming solid, gives firmness to the *crassamentum*, is proved by agitating fresh blood with a stick, so as to collect this substance on the stick, in which case the rest of the blood remains fluid *.

* It may be proper to mention here, that till of late the coagulable lymph has been confounded with the serum of the blood, which contains a substance that is likewise coagulable. But in these sheets, by the *lymph*, is always meant that part of the blood which jellies, or becomes solid spontaneously when blood is received into a bason, which the coagulable matter that is dissolved in the serum does not; but agrees more with the white of an egg, in remaining fluid when exposed to the air, and coagulating when exposed to heat, or when mixed with ardent spirits, or some other chemical substances.

THE

THE surface of the *craſſamentum*, when not covered with a ſize, is in general of a more florid red than the blood was when firſt taken from the vein, whilſt its bottom is of a dark colour, or blackiſh. This floridneſs of the ſurface is juſtly attributed by ſome of the more accurate obſervers to the air, with which it is in contact; for, if the *craſſamentum* be inverted, the colours are changed, at leaſt that which is now become the upper ſurface aſſumes a more florid redneſs. This difference of colour, others have endeavoured to explain from the different proportions of the red particles, or globules as they are called, which, ſay they, being in a greater proportion at the bottom of the *craſſamentum*, make it appear black; but, if inverted, the globules then ſettle from the ſurface which is now uppermoſt, and that becomes redder. But this I think is not probable; for the lymph in the *craſſamentum* is ſo firmly coagulated, as to make it too denſe to allow of bodies even heavier than the red

red particles to gravitate through it; for example gold. That air has the power of changing the colour of the blood, has been long known; and the following experiment shews it very satisfactorily, and hardly leaves room to refer the appearance to another cause.

EXPERIMENT III.

Having laid bare the jugular vein of a living rabbit, I tied it up in three places; then opening it between two of the ligatures, I let out the blood, and filled this part of the vein with air. After letting it rest a little, till the air should become warm, I took off the ligature which separated the air from the blood, and then gently mixed them, and I observed that the venous blood assumed a more florid redness, where it was in contact with the air-bubbles, whilst in other parts it remained of its natural colour.

There

THERE is a difference between the arterial and venous blood in colour; the former is of a florid red like the surface of the *craſſamentum*, the latter is dark or blackiſh like the bottom of the *craſſamentum*. This change in its colour is produced on the blood as it paſſes through the lungs, as we ſee by opening of living animals *; and as a ſimilar change is produced by air applied to blood out of the body, it is preſumed that the air in the lungs is the immediate cauſe of this change; but how it effects it, is not yet determined.

* That this change is really produced in the lungs, I am perſuaded from experiments, in which I have diſtinctly ſeen the blood of a more florid red in the left auricle, than it was in the right. But ſome authors of the greateſt authority ſay, that they could not obſerve any ſuch difference in a great number of experiments which they made; but this I ſhould attribute to their having been later in opening the left auricle after the collapſing of the lungs than I was; for it ſeems probable, that whatever is the alteration produced on the blood in its circulation through this organ, that change cannot take place after it is collapſed.

As the blood is changed to a more florid red in paffing through the lungs, or from the venous to the arterial fyftem, fo it lofes that colour again in paffing from the arteries to the veins in the extreme parts, especially when the perfon is in health; but every now and then we obferve the blood in the veins more florid than is ufual, and it likewife frequently happens in venefection, that the blood which comes firft out is blackifh, and that which comes afterwards is more florid: in fuch cafes, the arterial blood paffes into the veins without undergoing that change which is natural to it.

Some of the neutral falts have a fimilar effect on the colour of the blood to what air has, particularly nitre; thence fome have attributed the difference of colour in the arterial and venous blood to nitre, which they fuppofed was abforbed from the air whilft in the lungs. But we know that this is a mere hypothefis, for air contains no nitre. Indeed

nitre

nitre is far from being the only neutral salt which has this effect on the blood, for moſt of them have ſome degree of it. In making ſome experiments on this ſubject, I have obſerved a more remarkable effect which ſome of the neutral ſalts have upon the blood; and that is, being mixed with it when juſt received from the vein, they prevent its coagulation, or keep it fluid, and yet, upon adding water to the mixture, it then jellies or coagulates. Thus, if ſix ounces of human blood be received from a vein upon half an ounce of true Glauber's ſalt reduced to a powder, and the mixture agitated ſo as to make the ſalt be diſſolved, that blood will not coagulate on being expoſed to the air, as it would have done without the ſalt; but if to this mixture about twice its quantity of water be added, in a ſhort time the whole will be jellied or coagulated, and on ſhaking the jelly, the *coagulum* will be broken, and the part ſo coagulated can

can be now separated as it falls to the bottom, and proves to be the lymph.

In these mixtures of the blood with neutral salts, the red particles readily subside, (especially if human blood be used) and the surface of the mixture becomes clear and colourless; and being poured off from the red part, it is found to contain the coagulable lymph, which can be separated by the addition of water.

I have tried most of the neutral salts, and have made a table of their effects on the blood; but it is not necessary to trouble the reader with it, as we do not see of what use it could be in medicine; for we must not conclude that their effects within the body would be the same as out of it *. Indeed, these experiments,

* The salts which keep the blood fluid by itself, and yet allow it afterwards to jelly on being mixed with water, are, *Sal Glauberi verus*; *Sal Digestivus Sylvii*,

periments, as well as some others, were not made so much with a view of any immediate application to medicine, as to determine the properties of the blood chemically; for, having set out with a persuasion, that a more particular acquaintance with the properties of this fluid was necessary before we could arrive at the knowledge of some of the animal functions, such as the manner in which the bile and other secreted fluids are formed, I therefore was anxious to throw some more light on this subject. With this view I have made some experiments even on living animals, being convinced

Sylvii; Sal Communis; Nitrum Commune; ——— Nitrum Cubicum; Sal Diureticus; Borax. The salt made of Vinegar and the fossile Alkali; and the salt made with Vinegar and Chalk.

The following salts likewise keep the blood fluid, but do not allow it to jelly when mixed with water; Tartarus Vitriolatus; Sal Epsomensis; Sal Ammoniacus Communis; Sal Ammoniacus Nitrosus; Sal Rupillensis:

But Alum on being mixed with blood coagulates it immediately.

that my inquiries would not otherwise be satisfactory.

When blood is thus kept fluid by Glauber's salt, it still retains its property of being coagulable by heat, and by other substances as before, air excepted. This method of keeping the blood fluid may therefore be useful, by affording an opportunity of making some experiments upon it, which we could not otherwise do from its coagulating so soon when taken from the vessels.

This property of one of the neutral salts has been long known amongst those who prepare the blood of cattle for food; for it has long been a practice with such people, to receive it into a vessel containing some common salt, and to agitate it as fast as it falls, by which means the coagulation is prevented, and the blood remains so fluid as to pass through a strainer, without leaving any *coagulum* behind:

behind: by this means they have an opportunity of mixing it with other fubftances for culinary purpofes.

Although the coagulable lymph fo readily becomes folid when expofed to the air, yet whilft circulating it is far from that confiftence: it has indeed been fuppofed to be fibrous, even whilft moving in the blood-veffels, but erroneoufly.

It is this coagulable lymph which forms the inflammatory cruft, or *buff*, as it is called. It likewife forms *polypi* of the heart, and fometimes fills up the cavities of aneurifms, and plugs up the extremities of divided arteries. It is fuppofed, by its becoming folid in the body, to occafion obftructions and inflammations; and even mortifications, from the expofition to cold, have been attributed to its coagulation. In a word, this lymph is fuppofed to have fo great a fhare in the caufe of feveral difeafes, that it would be a defirable matter to be able to afcertain the caufes of that coagulation, either in the body, or out of it.

THE blood, when received into a bafon, and fuffered to reft in the common heat of the atmofphere, very foon jellies or coagulates; the part which now becomes folid is the coagulable lymph, as has been fhewn above. The circumftances in which it now differs from what it was in the veins, are thefe: it is expofed to the air, to cold, and is at reft; for whilft in the body, air is excluded, it is there of a confiderable warmth, and is always in motion. The queftion is, to which of thefe circumftances its coagulation whilft in the bafon is chiefly owing. This queftion, I believe, cannot well be anfwered from the experiments that have hitherto been made. It has indeed been faid, that the cold alone coagulated it; for, fay they, if you receive blood into a bafon, and fet that

that bason in warm water, and stir the blood well, it can be kept fluid. But in the experiments from which this conclusion was made, I find there has been a deception *. In short, I have found that it coagulates as soon when kept warm and when agitated, as it does when suffered to rest and to cool. As the subject seemed to me of importance, I have endeavoured to ascertain the circumstance to which this coagulation is owing by several experiments, in each of which the blood was generally exposed to but one of the suspected causes at a time. Thus, in order to see whether the blood's coagulation out of the body was owing to its being at rest, I made the following experiment:

* That is, the lymph really had been coagulated, but by the agitation had likewise been separated from the rest of the blood, and had thereby escaped notice.

EXPERIMENT IV.

Having laid bare the jugular vein of a living dog, I made a ligature upon it in two places, so that the blood was at rest between the ligatures; then covering the vein with the skin, to prevent its cooling, I left it in this situation. From several experiments made in this way, I found in general, that after being at rest for ten minutes, the blood continued fluid; nay, that after being at rest for three hours and a quarter, above two-thirds of it were still fluid, though it coagulated afterwards. Now the blood, when taken from a vein of the same animal, was completely jellied in about seven minutes. The coagulation therefore of the blood in the bason, and of that which is merely at rest, are so different, that rest alone cannot be supposed to be the cause of the coagulation out of the body.

the Properties of the Blood. 19

To see the effects of cold on the blood, I made this experiment:

EXPERIMENT V.

I KILLED a rabbit, and immediately cut out one of its jugular veins, proper ligatures being previously made upon it; I then threw the vein into a solution of sal ammoniac and snow, in which the mercury stood at the 14th degree of Fahrenheit's thermometer. As soon as the blood was frozen and converted into ice, I took the vein out again, and put it into lukewarm water till it thawed and became soft; I then opened the vein, received the blood into a tea-cup, and observed that it was perfectly fluid, and in a few minutes it jellied or coagulated as blood usually does. Now, as in this experiment the blood was frozen and thawed again without being coagulated, it is evident that the coagulation of the blood out of the body is not solely owing to cold, any more than it is to rest.

Next, to fee the effects of air upon the blood, I tried as follows:

EXPERIMENT VI.

Having laid bare the jugular vein of a living rabbit, I tied it up in three places, and then opened it between two of the ligatures, and emptied that part of its blood. I next blew warm air into the empty vein, and put another ligature upon it, and letting it reft till I thought the air had acquired the fame degree of heat as the blood, I then removed the intermediate ligature, and mixed the air with the blood. The air immediately made the blood florid, where it was in contact with it, as could be feen through the coats of the vein. In a quarter of an hour I opened the vein, and found the blood entirely coagulated; and as the blood could not in this time have been completely congealed by reft alone, the air was probably the caufe of its coagulation.

From comparing thefe experiments, may we not venture to conclude, that the air is a ftrong coagulant of the blood, and that to this its coagulation when taken from the veins is chiefly owing, and not to cold, nor to reft?

But although it appears from thefe experiments, that the coagulation of the blood in the bafon is owing to the air alone; for cold has no fuch effect, nor has reft in a fufficient degree, becaufe the coagulation of the blood in the bafon takes place in a few minutes, whilft that which is merely at reft in the veins is not completely coagulated in three hours or more. Yet the blood is in time completely coagulated merely by its being at reft in the veins; but then in this cafe it coagulates in a different manner from what it does in the bafon; and as it probably is in this way that the blood is coagulated in the body, I have been more particularly attentive to it, and have endeavoured to determine by experiment how it takes place. With this view I have feveral times repeated Experiment the 4th, which was made with a view to determine whether the blood would coagulate by reft. In the firft trial, the vein was not opened till the end of three hours and a quarter; and juft before it was opened I had obferved through its coats, that the upper part of the blood was tranfparent, owing to the feparation of the lymph. On letting out this blood, it feemed to me entirely fluid: a part indeed had been loft, but the greateft part was collected in the cup, and which afterwards coagulated as blood commonly does when expofed to the air. From this experiment I imagined that the whole had been fluid; but from others made fince, I am perfuaded that the part which was loft had been coagulated; for, from a variety of trials, I now find, that though the whole of the blood is not congealed in this time by reft alone, yet a part of it is. But as it would be trefpaffing too much

much on the reader's time to relate every experiment I have been obliged to make for this purpofe, I fhall only mention the general refult of the whole.

AFTER fixing a dog down to a table, and tying up his jugular veins, I have in general found, that on opening them at the end of ten minutes, the blood was ftill entirely fluid, or without any appearance of coagulation *. If they were opened at the end of fifteen minutes, at firft fight it alfo appeared quite fluid; but on a careful examination I found fometimes one, and fometimes two or three fmall particles about the fize of a pin's head, which were coagulated parts of the blood. When opened later than this period, a larger and larger

* I fay *in general* it was fluid at the end of ten minutes; but I muft likewife mention that in one dog I found two very fmall particles of beginning coagulation, even at this period; yet in another I could not obferve any fuch appearance, even at the end of fifteen minutes.

coagulum was obferved; but fo very flowly does this coagulation proceed, that in an experiment where I had the curiofity to compare more exactly the clotted part with the unclotted, I found, after the vein had been tied two hours and a quarter, that the *coagulum* weighed only two grains; whilft the reft of the blood, which was fluid, on being fuffered to congeal, weighed eleven grains. I can advance nothing farther in this part of my fubject with precifion. Nor can I pretend exactly to determine the time at which all the blood between the ligatures is coagulated. I have indeed opened fuch a vein at the end of three days, when I found a thin, white *coagulum*, which was a mere film; the *ferum* and red particles having difappeared. But the whole is undoubtedly congealed long before this period. The manner in which the blood coagulates, when at reft in the body, has appeared to me curious, and therefore I have taken the more pains to difcover how it happens, efpecially

cially as it may affift us in judging whether or no it coagulates in the heart, fo as to form thofe fubftances called *polypi*. The abovementioned times will, I believe, be found to be thofe at which the blood congeals in the veins of healthy dogs: and as I have found, by experiments, that the blood of a dog and of the human fubject in health jellies out of the body, nearly in the fame time, that is, it begins in three or four minutes, and is completed in feven or eight; I fhould therefore conclude that the blood in the veins of the human body coagulates nearly at the fame period with that of a dog. But it may be neceffary to add here, that from experiments which I have made, I have reafon to believe that the time at which the blood coagulates, is different in different conftitutions, and in different difeafes. For though the blood of a perfon in health is completely coagulated in feven minutes after it is taken out of the veins, yet in fome difeafes, I have found the blood fifteen or twenty

twenty minutes, nay even an hour and an half, before it was completely jellied.

As we fee in the above-related experiments, that the blood coagulates in the body when fuffered to reft for a little time, is it not probable that to this caufe its coagulation in thofe true aneuryfms, which are attended with a pouch, are owing *? For in fuch enlargements a part of the blood is without motion, which will congeal when at reft, and in contact with the fack; and thus one layer may be formed; and the fack afterwards enlarging, another portion of the blood will then be at reft; and fo a fecond layer may be formed; and thence probably is the origin of thofe laminated *coagula* met with in fuch facks.

Likewise, to the blood's being at reft, is probably owing its coagulation in the

---

* An inftance of which may be feen in the Med. Obf. and Inq. vol. i. article xxvii. fig. iii.

large

large arteries which are tied after amputation, or other operations; for after moſt of ſuch ligatures there will be a part of the artery impervious, in which the blood can have no motion. The *coagulum* after amputation might indeed be ſuppoſed owing to air; but, conſidering the manner in which arteries are tied whilſt the blood is running from them, it does not ſeem probable that the air has any effect on what is above the ligature.

To the blood's being without motion in the cavity of the *uterus*, is its coagulation therein probably owing; hence the origin of thoſe large clots which ſometimes come from that cavity; and which, when more condenſed by the oozing out of the *ſerum*, and of the red globules, aſſume a fleſh-like appearance, and have often been called *Moles* or *falſe conceptions*.

In

In Experiment the 5th, we found that the blood could be frozen and thawed again, without being coagulated: this likewise is an experiment which I have repeated several times, that I might be sure of the fact. I have also varied the experiment by sometimes putting the vein into a phial of water, and freezing the whole in a solution of sal ammoniac in snow; and sometimes I have put the vein into the solution itself; and three or four times I have thrown it into oil, and then frozen it; but after all these trials, the result was the same. The blood was always evidently fluid on being thawed, and as evidently jellied when exposed to the air.

Besides being coagulated when exposed to the air, the coagulable lymph, as well as the *serum*, is known to be fixed by heat; but the degree of heat has not, so far as I know, been determined. It has been supposed to require a degree almost

almoſt equal to that which coagulates the *ſerum* *; but one much leſs is neceſſary, as will appear from the following experiments.

EXPERIMENT VII.

HAVING found, from a number of trials, that blood, kept fluid by means of true Glauber's ſalt, had its lymph coagulated by a heat of 125° of Fahrenheit's thermometer, I ſuppoſed that the degree neceſſary for fixing it in its natural ſtate could not be very different from this. I therefore prepared a lamp-furnace with a ſmall veſſel of water upon it; this water was heated to 125°; and then laying bare the jugular vein of a living dog, I tied it properly, cut a piece of it out, and put it into this water: after eleven minutes, I took out the

---

* See Traité du Cœur. T. ii. p. 93. Schwenke Hæmatolog. p. 138.

vein,

vein, opened it, and found the blood entirely coagulated; thence I concluded, that 125°, or less, was sufficient to coagulate the blood of a dog. It may be necessary to observe here, that the part coagulated was only the lymph; for the *serum* requires a much greater heat to fix it, that is a heat of 160°, as will appear hereafter.

EXPERIMENT VIII.

THE same experiment was repeated in such a manner, that the heat never went higher than 120° and an half; and I found, that on opening the vein at the end of eleven minutes, the lymph was entirely coagulated, even in this heat.

EXPERIMENT IX.

I REPEATED the experiment with a heat no higher than 114°, and at the end of eleven minutes, the vein being opened, the

the blood was found to be fluid, and in a few minutes after, being laid open to the air, it coagulated as usual. Now, as the blood in the last experiment was coagulated, when the heat had never risen above 120° and an half; and in this experiment was fluid, though it had been exposed to a heat of 114°; we may therefore conclude, that the coagulable lymph in the blood of a dog, in health, is fixed in a degree of heat between 114° and 120°$\frac{1}{2}$ of Fahrenheit's thermometer.

As to the degree of heat at which the lymph in human blood coagulates, I have not yet had an opportunity of trying it in a more satisfactory way, than with the mixture with Glauber's salt, in which state it coagulates at 125°. But, as we find that the human blood and that of a dog jelly nearly in the same time, when exposed to the air, I think it probable that the precise degree of heat at which the lymph of the human blood coagulates,

lates, is between 114° and 120°½. I have thought of making the experiment on the umbilical cord of a recent *placenta*, which perhaps is the moſt likely way of coming at the truth.

THE degreé of heat, at which the *ſerum* of the blood (which ſhould not be confounded with the lymph) coagulates, is generally ſaid to be 150°; but from my experiments I am inclined to believe it requires a greater heat to fix it. They were made in the following manner:

EXPERIMENT X.

I TOOK a wide-mouthed phial, containing ſome *ſerum*, and placing a thermometer in it, I put it into water which was kept warm by a lamp underneath; and, in making this experiment with as much accuracy as I could, I found the heat required was 160°; which is about
forty

forty degrees more than is neceffary for the coagulation of the lymph.

As the blood is coagulable by heat, and as the heat of an animal is increafed in fevers, it has been fuppofed that the blood might be coagulated by the animal heat, even whilft it is circulating in the veffels; but there is little foundation for fuch an opinion, fince the animal heat is naturally only 98° or 100°, and in the moft ardent fever is not raifed above 112°.

CHAP. II.

Of the inflammatory Cruft, or Size.

I SHALL next proceed to inquire into the formation of the inflammatory cruft, or *fize*, as it is called.

This remarkable appearance is frequently met with in inflammatory diforders, and is formed by the coagulable lymph's being fixed, or coagulated, after the red particles have fubfided. It has indeed been fuppofed to be formed from the *ferum* of the blood; and an excellent writer on this fubject feems in doubt to which of the two it fhould be attributed. But that it is formed by the coagulable lymph alone, after the red particles have fubfided, appears from the following experiments.

EXPERIMENT XI.

In the month of June, when the thermometer in the shade stood at 67.°, I bled a man who had laboured under a *phthisis pulmonalis* for some months, and at that time complained of a pain in his side. The blood, though it came out in a small stream, yet flowed with such velocity, that it soon filled the bason. After tying up his arm I attended to the blood, and observed that the surface became transparent, and that the transparency gradually went deeper and deeper, the blood being still fluid. I likewise observed that the coagulation first began on the surface, where it was in contact with the air, and formed a thin pellicle; this I removed, and saw that it was soon succeeded by a second. I then took up a part of the clear liquor with a wet teaspoon, and put it into a phial with an equal quantity of water; a second portion

I kept in the tea-spoon; and I found afterwards that they both jellied or coagulated, as did also the surface of the *crassamentum*, making a thick crust. On pressing with my finger that portion which was in the tea-spoon, I found it contained a little *serum*.

From this experiment it is evident, that the substance which formed the size was fluid after it was taken from the vein, and coagulated when exposed to the air; and as this is a property of the coagulable lymph alone, and not of the *serum*, there can be no doubt that the size was formed of the lymph.

The following experiment, made on the blood, without exposing it to the air, likewise proves the same fact.

EXPERIMENT XII.

IMMEDIATELY after killing a dog, I tied up his jugular veins near the *sternum*, and hung his head over the edge of the table, so that the parts of the veins where the ligatures were might be higher than his head. I looked at the veins from time to time, and observed that they became transparent at their upper part, the red particles subsiding. I then made a ligature upon one vein, so as to divide the transparent from the red portion of the blood; and opening the vein, I let out the transparent portion, which was still fluid, but coagulated soon after. On pressing this *coagulum*, I found it contain a little *serum*. The other vein I did not open till after the blood was congealed, and then I found the upper part of the *coagulum* whitish like the crust in pleuritic blood *.

AND that the size is merely the coagulable lymph separated by the subsidence of the red particles, will appear evident to any person who will, as Sydenham directs, move a finger, or a tea-spoon through the blood when he observes its surface becoming transparent; for in this case the blood that otherwise would have been sizy, will now have a natural appearance, or be without size; from the red particles being prevented from subsiding.

---

* This is not the only apparently healthful animal whose blood had a crust; I have seen it in others: whence I at first suspected that merely keeping the blood fluid for a little time was sufficient to produce this appearance; but I altered my opinion, on seeing, that in the greatest number of animals it did not occur: nor is it commonly met with in the hearts of those persons who die a violent death, though the blood remains longer fluid in such cases, than it does in the bason where the size appears.

It has been a very generally received opinion, that inflammation thickens the blood, and makes it more ready to coagulate. Nay, some have gone so far as to say, that in those disorders where the inflammatory crust is seen, the blood is almost coagulated even before it is let out of the vein. Now I am persuaded from experiment, that the contrary of this is true; or that inflammation, instead of increasing the disposition of the blood to coagulate, really lessens it; and instead of thickening the blood, really thins it; at least that part which forms the crust, viz. the coagulable lymph.

In the first place, that inflammation really lessens the disposition to coagulate, will appear evident to every one who attends to the jellying of such blood as has a crust. For in all those cases the blood will be found to be longer in congealing, than in its natural state. To this opinion I was first led by attending to the phthisical patient's blood abovementioned; but

I have since made a comparison, which seems to prove the fact. For, from a variety of experiments made on the blood of persons nearly in health, or at least who had no inflammatory disorder, and no crust on their blood, I found that after being taken from a vein, it began to jelly in about three minutes and an half. The first appearance of coagulation was a thin film on the surface near the air-bubbles, or near the edge of the bason; this film spread over the surface, and thickened gradually till the whole was jellied, which was in about seven minutes after the opening of the vein; and in about ten or eleven the whole was so firm, that, on cutting the cake the gashes were immediately filled up by the *serum*, which now began to separate from the *crassamentum*. But in those persons whose blood had an inflammatory crust, the coagulation was much later; as will appear from the following experiments.

EXPERIMENT XIII.

I bled a woman who was seven months gone with child, and the blood was received into a bason. In five minutes after the vein was opened, a film first appeared; but this spread so slowly, that in ten minutes it did not cover the whole surface; in fifteen minutes it had nearly spread over the surface; but the rest of the blood was quite fluid, at least for some depth, and even in half an hour it was not so firmly jellied as it was afterwards. In this case there was a very thick and strong crust or size.

EXPERIMENT XIV.

Having bled a person with a violent rheumatic pain in his breast, the blood was received into three tea-cups, and each

each of them had afterwards a cruſt. In the firſt I obſerved the progreſs of the coagulation, as follows: The beginning of the coagulation was not marked, but at the end of half an hour the film was not thicker than common writing-paper; and this being removed, a little of the clear lymph was taken up with a wet tea-ſpoon, put into a clean cup, and was twenty minutes more in coagulating. Even at the end of an hour and an half, the whole of the blood was not jellied; for at this time I removed the film or pellicle, and took up a ſecond portion of clear lymph with a ſpoon, and put it into a tea-cup, where it jellied afterwards; though this jelly was not indeed quite ſo firm as the *craſſamentum* itſelf.

EXPERIMENT XV.

A WOMAN, with a ſlight inflammation in her throat, had eight ounces of blood taken from her arm; the blood was received

received into a bafon, and the bleeding finifhed in four minutes and three quarters, when a film began to form near the air-bubbles; in feven minutes a tranfparent fize appeared over a confiderable part of the furface which was quite fluid, whilft the reft of the blood was coagulating, there being now a very diftinct red cruft over the reft of the furface.

Now, from comparing thefe experiments with what has been obferved of the coagulation of the blood, where there is no inflammatory cruft or fize, is it not evident that the blood remains longer fluid after being expofed to the air, and has lefs difpofition to coagulate, in thofe cafes where there is a fize, than where there is none? for where there was none, it was found to coagulate completely in feven minutes; but in one of the others, where the fize was very thick, it did not completely coagulate in lefs than an hour and an half.

THE power that inflammation has in leffening the difpofition of the lymph to coagulate is likewife plain from the following experiment, where the blood in the heart of a dead animal feems to have congealed very flowly.

EXPERIMENT XVI.

A DOG was killed eight hours after receiving a large wound in his neck. The wound had during this time inflamed confiderably. Upon opening him next morning, when he had been dead thirteen hours, a large whitifh *polypus* was found in the right ventricle of his heart; under this was a little blood ftill fluid, which being taken up with a tea-fpoon, coagulated foon after being expofed to the air.

IT may be proper to obferve here, that in the hearts of animals which had died without any inflammation, I have found the

the blood entirely coagulated long before this time. And that from opening them at different times, I have feen it coagulate in their hearts after death, in the fame gradual manner that it does in their veins, when its motion is ftopt by ligatures; as related in page 23.

In the next place, that the blood is really attenuated in inflammatory diforders, where the whitifh cruft or fize appears, is probable from the following circumftances: Firft, It even feems thinner to the eye; 2dly, The red particles or globules fubfide fooner in fuch blood, than in that of an animal in health. This feems proved by obferving that in the above-mentioned experiments, where the blood was at reft in the veins, it was not covered with a cruft, except in one or two inftances, though in all thofe cafes it remained longer fluid than the blood commonly does in a bafon, after bleeding, where the cruft appears. And again, the blood in the heart of an animal

mal that dies a violent death, is not generally covered with a white cruft, notwithſtanding it is ſo late in being congealed. Theſe circumſtances ſhew, that ſomething more than merely a leſſened difpoſition to coagulate is neceſſary for the forming of the cruft or ſize. 3dly, The globules more readily ſubſide in inflammatory caſes, from the ſurface of the whole maſs of blood, than they will afterwards do from the ſurface of a mixture with the *ſerum* alone, of which the following experiments are a proof. But, before I relate them, let me obſerve, that they were made with a view to diſcover, whether the inflammatory cruft could be owing to any other cauſe than to the attenuation of the coagulable lymph, and to its difpoſition to coagulation being leſſened: and as the ſame appearance might be ſuſpected to ariſe from an increaſed ſpecific gravity in the red particles, or from the *ſerum* alone being attenuated, I endeavoured to decide the queſtion in the following manner.

EXPE-

EXPERIMENT XVII.

INTO a phial, marked A, I put an ounce of the *serum* of the blood of a person, whose *crassamentum* had an inflammatory crust.

INTO another, marked B, I poured an ounce of the *serum* of a person whose blood had no crust; then to each of these, I added a tea-spoonful of *serum*, loaded with the red particles of a person whose blood had no inflammatory crust or size. In attending to them, I could not observe that the red particles subsided at all sooner in the *serum* of the blood that had a crust, than they did in the *serum* of that blood which had no crust. Thence I conclude, that the *serum* is not attenuated in those cases where the inflammatory crust appears.

LASTLY,

LASTLY, To see whether the specific gravity of the red globules was increased, I proceeded as follows:

EXPERIMENT XVIII.

I POURED into a phial C, a portion of the *serum* of the blood which had no crust; and likewise into another D, a second portion of the same *serum*. I then added to C a tea-spoonful of the same *serum*, loaded with red particles from the blood which had an inflammatory crust. And into D I poured a tea-spoonful of the same *serum*, loaded with the globules of that blood which had no crust. In viewing these, I could not observe that the globules of the blood which had an inflammatory crust subsided sooner than those of the blood which had none: whence I inferred, that the specific gravity of the red particles, or globules as they are called, is not increased in those
cases

cafes where the cruft appears. And, therefore, fince that inflammatory cruft or fize feems neither owing to the *ferum*'s being attenuated, nor to an increafed fpecific gravity in the red particles, it probably depends folely upon a change in the coagulable lymph. And what feems farther to confirm this inference, in none of thefe experiments did the red particles fubfide from the furface of the *ferum* in 20 minutes, though, where the cruft appears, they fubfide from the furface of the blood in half that time; fo that the whole mafs of blood feems to be thinner than the *ferum* alone; or, the coagulable lymph feems to be fo much attenuated in thefe cafes, as even to dilute the *ferum*, which at firft fight appears a paradox.

MAY we not, therefore, conclude, that in thofe cafes where the inflammatory cruft appears, the coagulable lymph is thinner, and its difpofition to coagulation is leffened? both of which circumftances contribute

contribute to the subsiding of the red globules from the surface of the blood, which then coagulating gives rise to this appearance, called the inflammatory crust or size, in the blood of pleuritic or rheumatic patients *.

How contrary to the conclusion, which these experiments lead us to, are the opinions of some medical writers on this subject! How frequently do we find it said, that the blood is thicker in inflammatory disorders, where that size occurs; and that a large orifice is necessary to let out the vitiated blood! That a large orifice is preferable to a small one in many cases, where such blood is found, I believe to be true; but that its advantages are owing to its letting out the thickened blood, seems improbable from what we

* This remarkable appearance might indeed be accounted for, by supposing that the lymph had ascended to the surface of the blood in those cases; but this is improbable, from considering, that, in its coagulated state, it is of greater specific gravity than the serum, and sinks in it.

have feen in the experiments above related: they are perhaps nearer the truth, who attribute it to the fuddenneſs of the evacuation.

It may be proper to obſerve here, that this fize or whitiſh cruſt is not a certain ſign of inflammation; it being often met with where there ſeems to be no ſuch diſeaſe, in particular in the blood of pregnant women. And that it differs much in denſity in different caſes; in ſome it is extremely firm, in others it is ſpungy or cellular, and contains much *ſerum* in its cells. Theſe diverſities we ſhall endeavour to explain hereafter, when we have laid before the reader ſome more obſervations on the coagulation of the lymph †.

† Although this Eſſay has been ſo lately printed, yet moſt of the facts which occur in the preceding pages have been mentioned in my Anatomical Lectures, ever ſince the year 1767; and ſome of them were mentioned publicly even before that time. This I thought neceſſary to obſerve, becauſe many of them have ſince appeared in other publications.

CHAP. III.

Of the causes of the inflammatory crust's appearing at different times in blood-letting; of the stopping of hemorrhages; and of the effects of cold upon the blood.

IT has been observed by those who have written on the blood, that it sometimes happens in blood-letting, that the first cup has an inflammatory crust, whilst the last has none; but no satisfactory reason has been given for this difference. One might suppose that it was owing to some circumstance in the bleeding, such as in the different velocity with which the blood flowed into each cup, or to the last cup's being agitated so as to prevent the separation of the lymph: but I have seen it where there

was

was no difference of this fort, nor in any other circumſtance that I could obſerve. I therefore ſuſpect that in ſuch caſes the properties of the blood are changed, even during the time of the evacuation; to which opinion I was led by the following experiments.

EXPERIMENT XIX.

NINE ounces of blood were taken from a woman who had been delivered two days before, and who at that time laboured under a fever, with a conſiderable pain in her ſide, and in her *abdomen*. The blood was received into a baſon, and her arm was tied up; when, on looking at the blood, I found its ſurface tranſparent for ſome depth, an indication of a future cruſt; and as her pain was not abated, and as her pulſe could bear it well, I removed the ligature from her arm, and took away about ſix ounces more, into three tea-cups; but what appeared

appeared to me remarkable, although the blood flowed as faſt into each of the cups as into the baſon, and when full they were immediately ſet down on the ſame window, yet there was no inflammatory cruſt on the blood in the cups, though a very denſe one on that in the baſon. And again, although the blood in the baſon had been taken away ſome minutes before that in the cups, yet it was later in being completely coagulated; as was evident on comparing them.

I HAD an opportunity of repeating the experiment in the evening; for the ſymptoms of inflammation ſeeming equally violent, it was thought proper by the phyſicians who attended her, to take away more blood; which was done by opening the ſame orifice, when three tea-cups were nearly filled, and ſet in the ſame place; and it was obſerved, that the firſt had a cruſt, though not ſo thick a one as in the firſt bleeding; but the other

other two cups were without this appearance, though the blood had flowed into them even more quickly than into the first †.

EXPERIMENT XX.

A GENTLEMAN, who laboured under an inflammatory complaint, had about nine ounces of blood taken from his arm.

---

† As this experiment seems contradictory to some mentioned hereafter, in the last cups being filled rather sooner and yet coagulating sooner, which might be suspected to be owing to the vessel's acting more strongly at the latter part of the operation than at the beginning; it is therefore necessary to observe, that the difference in this experiment appeared to be only owing to a difference in the size of the orifice; for when the ligature was first removed, the old wound was not so much torn open as it was afterwards, when it was more enlarged in order to hasten the evacuation. But it did not, in the beginning of the operation, trickle down the arm as in Experiment 27; where the size of the orifice was not enlarged from the first, and yet in proportion as the operation advanced, the velocity of the blood increased; which was thence concluded to be owing to an increased action of the blood-vessels.

This quantity was divided into four portions; the firſt was received into a cup, and was in meaſure little more than an ounce; the ſecond, into a baſon, to the quantity of two ounces; the third into a cup, which held one ounce; and the fourth into a baſon, to the quantity of three ounces. Each veſſel was immediately placed upon the window; and it was obſerved that the blood in the firſt was lateſt in coagulating, and had a cruſt over the whole ſurface; that in the ſecond had a cruſt only upon a part of its ſurface; but that in the third and fourth had none, and manifeſtly coagulated before either of the other two.

Now, ſince in theſe experiments the blood in the firſt cups was later in coagulating than that in the laſt, and ſince the blood in the firſt cups alone had a ſize, is it not probable, that even during the ſhort time taken up in the evacuation, the properties of the lymph had been changed,

changed, and that it was owing to this change that the size disappeared? It might indeed, at first sight, seem possible, that the bleeding had only let out the vitiated part; but this is not at all likely; for, suppose a part only of the blood was vitiated, that part must have been equally diffused through the whole mass, and there is no probability of its getting out of the vessels before the rest of the blood; and consequently it ought to have appeared in the last equally as in the first cup, but it did not. Bleeding, therefore, in those cases alters the nature of the blood, not by removing the vitiated part, and giving room for new blood to be formed, as has been supposed; but probably by changing that state of the blood-vessels, on which the thinness, and lessened tendency of the lymph to coagulation, depends; which surely is a very curious circumstance ‡.

FROM

‡ That the properties of the blood can be changed by emptying the blood-vessels, is likewise proved by

From this obfervation we may be led to think, that it may be ufeful to receive the blood more frequently into fmall cups, inftead of a bafon, and to attend more carefully to the alteration produced upon it by bleeding; as we may by that means perhaps learn to determine better, what quantities fhould be taken away in particular cafes. For it would feem probable that the operation is likely to have the moft effect on the difeafe, in thofe cafes where the greateft change is produced by its means, on the difpofition of the blood to coagulate; and of that change, we can judge, by comparing the blood in the firft cup, with that in

an experiment hereafter to be related; where the blood in an animal in health was found to have its difpofition to coagulation increafed, in proportion as the veffels were emptied, and as the animal became weaker. It may likewife be proper to mention, that though the inference is here drawn from two experiments only, yet I have likewife obferved the fame appearance in other cafes, which I have thought unneceffary to relate.

the

the laſt; for the firſt cup will nearly ſhew the ſtate of the blood at the beginning; and the laſt cup the ſtate of the blood at the latter part of the evacuation.

It frequently happens, that inſtead of an inflammatory cruſt over the whole ſurface of the *craſſamentum*, there is only a partial one, which appears in large ſpots or ſtreaks. In ſuch caſes I have obſerved, that only a part of the blood had its diſpoſition to coagulate leſſened, as in Experiment XV. in which ſome of the blood remained fluid and tranſparent, where thoſe ſtreaks appeared, for ſome time after the coagulation had begun in other parts of the ſurface. Now whether in thoſe caſes there had been the ſame difference before the vein was opened, or whether the whole blood had not been of the inflammatory kind, before veneſection, and a part of it was changed as it ran out, or as ſoon as the general fulneſs was diminiſhed, may be a queſtion; but the probability, I think, is is much in favour of its being changed during the time of the evacuation, from what was observed in the last experiments.

When I had observed that this disposition of the lymph to coagulate was increased by bleeding, or by weakening the action of the blood-vessels, I suspected that possibly in those cases where the body was very weak, the disposition to coagulate might be so much increased, that instead of being three or four minutes in beginning to do it, after it is let out of the veins (as is the case in people in health) it might coagulate in less time, or almost instantaneously; for I imagined, that unless this took place, we could hardly conceive how the blood should ever have time to coagulate in ruptured vessels, so as to stop hemorrhages, as it is believed to do. And upon this occasion I recollected a remark that I had heard, particularly from Dr. Hunter, which

which is, " That the faintnefs which
" comes on after hemorrhages, inftead
" of alarming the bye-ftanders, and
" making them fupport the patient by
" ftimulating medicines, as fpirits of
" hartfhorn and cordials, fhould be
" looked upon as falutary; as it feems
" to be the method Nature takes to give
" the blood time to coagulate." Now
as this feemed to favour my fufpicion, I
determined to make the experiment.

EXPERIMENT XXI.

Believing it would be fufficient for this purpofe, to attend to the properties of the blood, as it flows at different times from an animal that is bleeding to death, I therefore went to the markets, and attended the killing of fheep; and having received the blood into cups, I found my notion verified. For I obferved, that the blood which came from the veffels immediately

diately on withdrawing the knife, was about two minutes in beginning to coagulate; and that the blood taken later, or as the animal became weaker, coagulated in lefs and lefs time; till at laft, when the animal became very weak, the blood, though quite fluid as it came from the veffels, yet had hardly been received into the cup before it congealed. I have alfo repeated the experiment, by receiving blood into different cups at different times, whilft the animal was bleeding to death; and though the time taken up in killing the animal was not commonly more than two minutes, yet I obferved, on comparing the cups, that the blood which iffued laft coagulated firft *. I have obferved likewife,

---

* It may be neceffary to mention a circumftance that has occurred in repeating thefe experiments; which is, that although the laft cup being taken from the animal when much reduced, always coagulated in lefs time than the firft; yet when four or five cups were ufed, the blood in them did not always coagulate precifely in the inverfe order of their being filled; for fometimes the fecond coagulated

the Properties of the Blood. 63

likewife, that the blood coagulates with a different appearance in proportion as the animal becomes weaker; that which follows the knife begins to coagulate in about two minutes; it firft forms a film or pellicle on the furface, which extends gradually through the whole blood, yet fo flowly that its progrefs may be obferved, efpecially if the pellicle be moved from time to time. But the blood that comes from the fainting animal is coagulated in an inftant, after it once begins. From this circumftance, that the difpofition of the blood to coagulate is increafed as the animal becomes weaker, we may draw an inference of fome ufe with regard to the ftopping of hemorrhages, viz. not to roufe the patient by

---

lated before the third. This circumftance at firft feemed contradictory to the general conclufion, but on a more careful examination, it was fufpected to be owing to the ftruggles (or temporary exertion of ftrength of the veffels) of the animal, and no difference was obferved in the expofition to cold or to air.

ftimulating

ſtimulating medicines, nor by motion, but to let that languor or faintneſs continue, ſince it is ſo favourable for that purpoſe; and alſo, that the medicines likely to be of ſervice in thoſe caſes, are ſuch as cool the body, leſſen the force of the circulation, and increaſe that languor or faintneſs *. For, in proportion as theſe effects are produced, the divided arteries become more capable of contracting, and the blood more readily coagulates; two circumſtances that ſeem to concur in cloſing the bleeding orifices †.

It

---

* Beſides giving ſtimulants and cordials to counteract the fainting, it is a common practice in many parts of England, to give women, who are flooding, conſiderable quantities of port-wine, on a ſuppoſition that it will do them ſervice by its aſtringency. But ſurely, from its increaſing the force of the circulation, it muſt be prejudicial in thoſe caſes. Perhaps many of the remedies called ſtyptics might be objected to for the ſame reaſon.

† It has of late been proved by experiments, particularly by thoſe of the ingenious Mr. Kirkland, that the larger arteries, when divided, contract ſo as

IT has been queftioned whether bloodletting can be properly recommended in hemorrhages, excepting in thofe that are attended with evident figns of *plethora*: but do not thefe experiments fhew, that a vein may be opened with propriety, even where there is no *plethora*, in order fuddenly to bring on weaknefs; by which the momentum of the blood may be fo diminifhed, and the difpofition of the lymph to coagulate may be fo increafed, as to ftop the hemorrhage? For, when we confider how foon the blood-veffels contract, and adapt themfelves to the quantity of blood which they contain, it feems to ftop the hemorrhage. But the large *coagula* which we fee in the orifices of the veffels of the *uterus* of thofe who die foon after delivery, and the ftopping of hemorrhages where the blood-veffels were ruptured on their fides and not entirely divided, make me believe that contracting the bleeding orifice is not the only method nature takes to ftop an hemorrhage. Her refources indeed are great, and fhe has often more methods than one of producing the fame effect.

not improbable that in some cases where the hemorrhage is not profuse, but long-continued, the strength of the patient may be so recruited, that the disposition to coagulate shall not be sufficiently increased, or the extremities of the vessels sufficiently contracted, for the stopping of the bleeding; but, by emptying the vessels suddenly, this effect may be obtained, and the hemorrhage may be stopt by the loss of less blood, than would have happened, had only the slow draining been continued.

Although the whitish crust so commonly seen in inflammatory disorders, has so very morbid an aspect, as might induce us to consider it as inflammatory, and to bleed repeatedly in all those cases where it occurs, yet I believe we should act improperly: for, to say nothing of pregnancy, in which the appearance is almost constant, there are few physicians that have not seen patients, who, even in fuch circumftances, were the worfe for this evacuation. Nor need we be furprifed that this fhould happen, confidering how foon in fome inftances this fize difappears; and if fo, may we not fuppofe, that it may likewife foon be formed, even by a fhort exertion of ftrength in the veffels? Perhaps this was the cafe in the gentleman mentioned in page 55, who in lefs than twenty-four hours after bleeding, had fymptoms of great weaknefs.

As it appears from thefe experiments, that the difpofition of the blood to coagulate is increafed by bleeding, it may be ufeful to attend more to this circumftance, and to compare the coagulation of the blood in the laft, with that in the firft cup, even in cafes that are not attended with the inflammatory cruft. And it may likewife be worth while to make the fame comparifon in thofe cafes where every cup has a cruft; which frequently

quently happens both in rheumatic and in phthifical complaints. By thefe means we may judge what effect the evacuation has produced on the ftrength or fulnefs of the veffels; and may perhaps, by infpecting the laft cup, efpecially if it contains only a fmall quantity, be able to guefs pretty nearly at the nature of the blood which remains in the body. In the rheumatic cafe mentioned in page 41, every cup contained this cruft; and although the blood in the laft cup coagulated in much lefs time than that in the firft, yet as it was later in coagulating than common, I fufpected what remained in the veffels had the fame difpofition; but the patient recovered without repeating the evacuation.

It may be mentioned here, that I have once or twice feen blood, which, when it firft began to coagulate, had on its furface a red pellicle, and underneath a tranfparent fluid, which afterwards formed

formed a cruft. In thefe cafes, if the red pellicle had not been removed before the reft of the blood had congealed, we might have concluded that no part of the blood had this difpofition to form a white cruft. This appearance, I fhould imagine, was owing to the blood, where in contact with the air, having coagulated before the red particles had time to fubfide, from that part of the lymph which had its difpofition to coagulation leffened.

The learned profeffor de Haen has taken notice of a curious appearance of the blood, which he could not account for; but which, I prefume, may be explained from fome of the above experiments. His obfervation is, " that, " having bled a perfon in a fever, the " blood was covered with an inflamma- " tory cruft, and upon examining the " *craffamentum* in one of the cups, he " found that it formed a fort of fack " contain-

"containing a clear fluid: this fluid
"being let out, and the cup set by, on
"examining it next morning, he ob-
"served a very firm crust covering the
"whole again, and extending to the
"bottom of the cup *." I once met with a case similar to this; for, having bled a person into four cups at ten o'clock in the morning, on looking at the blood afterwards, at five in the afternoon, I found the *serum* had not separated from the *crassamentum* in the first cup; but the *crassamentum* felt as if it contained a fluid in a bag, as professor de Haen has described it. Upon pressing it, the fluid gushed out, and in a few minutes after being exposed to the air, coagulated: there was however this difference in the two cases, that in mine the fluid was red, so that it formed a red crust over the first, which was white. Now this seems to have been owing to the blood's having first coagulated, where it

* Vide Rat. Medendi, cap. vi.

was in contact with the air and with the sides of the cup; and the fluid which gushed out was the *serum*, with a part of the coagulable lymph, which still remained fluid; but, when exposed to the air, it jellied or coagulated, as it naturally does. That one part of the lymph can remain fluid after the other is coagulated, is proved by some of the preceding experiments; and I have more than once seen blood, which appeared perfectly jellied soon after bleeding; yet, on cutting into the *coagulum*, a transparent fluid has oozed out, which afterwards jellied. And so slowly does this coagulation proceed in some cases, that, in an experiment mentioned before, a part of the blood in a dog's heart was found uncoagulated thirteen hours after death. And I have likewise distinctly observed, that in some cases where the disposition to coagulate was much lessened during the evacuation, the blood at the bottom of the cup has jellied, whilst the greatest part of the *size* at the top was yet fluid; there being only a thin pellicle on its furface, where it was in contact with the air.

ANOTHER inftance of a change in the properties of this coagulable lymph, which appears curious, was feen in fome experiments, where I had occafion to throw the blood into water, and into oil, during the winter feafon, whilft the heat of the water and of the oil was no greater than 41° of Fahrenheit's fcale. In all thofe experiments, I found that the difpofition to coagulate was leffened, the blood becoming more and more vifcid, but did not coagulate whilft in that degree of cold. I fhall next relate thofe experiments.

EXPERIMENT XXII.

THE jugular vein being properly tied, and then cut out from a rabbit juft killed, was thrown into water of 41° of heat,

and

and taken out at the end of half an hour; when the blood was found to be ſtill fluid, though rather more viſcid than natural; but, after being expoſed to the air, it coagulated.

EXPERIMENT XXIII.

Two pieces of the jugular vein of a dog, juſt killed, were put into water, in which the thermometer ſtood at 41°; one was taken out after twenty minutes, and the other after three quarters of an hour; the blood in both was found to be fluid, and to coagulate afterwards.

As it was evident from theſe experiments, that the water had leſſened the diſpoſition of the blood to coagulate, I next enquired to what property in the water this effect could be owing; and to ſee whether water that was warmer would not have the ſame effect, I made the following experiment.

EXPERIMENT XXIV.

ON December the thirteenth, I cut out two pieces of the jugular vein of another dog, immediately after his death. One piece was put into cold water, and the other into water kept warm by a lamp, fo that the heat never varied more than between 90 and 100°. At the end of three quarters of an hour, that in the warm water had in it a *coagulum* as large as a garden-pea; but that in the cold water, being let out into a cup, was quite fluid. Twenty minutes after being expofed to the air, that which had been in the cold water was coagulating; but that from the warm water neither then nor afterwards fhewed any figns of farther coagulation: fo that it feemed not only to have jellied whilft in warm water, but to have begun to part with its *ferum*. From this experiment, it feems probable that the *coldnefs* was that
property

property of the water to which the leſſened difpofition to coagulate was owing; but, to be more fure of this, and to fee whether the blood might not be kept fluid a longer time by thefe means, I tried as follows:

EXPERIMENT XXV.

On January the fourteenth, I cut out a piece of the jugular vein of another dog, and put it into oil, in which the thermometer ftood at 38°. At the end of fix hours it was taken out, and the red particles were obferved through the coats of the vein to have moftly fettled to one fide. The blood was let out into a cup, and was found to be fluid; at the end of fifteen minutes above one half was ftill fluid; in twenty-five minutes it feemed to be quite jellied. Now as in this experiment a fimilar effect was produced, as when the vein was put into water, it feems probable that it was the coldnefs coldness of the water, and of the oil, which had lessened the disposition of the lymph to coagulate.

EXPERIMENT XXVI.†

ANOTHER piece of the same vein was put into river-water, in which the thermometer stood at 38°, and was left till the next morning; when, after twenty-two hours and a quarter, it was taken out. The red particles did not seem to have subsided, as in the former experiment; but the vein being opened the blood was found to be fluid, though so viscid that it could barely drop from the vessel. The cup into which it was received was placed upon the window of a

---

† It is necessary to observe here, that great expedition should be used in making these experiments; for, unless the vein be cut out in a few minutes after the death of the animal, the experiment may not succeed, from the blood's having begun to coagulate.

moderately

moderately warm room, and was examined carefully from time to time; but the blood never had any appearance of coagulation, on the contrary, it remained fluid till it was dried by the evaporation of the water, which happened by the next day. In this experiment the cold feemed entirely to have prevented the coagulation of the lymph: fo ill-founded is the common opinion, that cold coagulates the blood.

As the lymph, on being cooled, is deprived of its power of coagulating when expofed to the air, may we not thence be led to explain that fact mentioned by Lifter, that the blood of thofe cold animals which fleep during the winter-feafon, on being let out into a bafon, does not coagulate? And thence, as he obferves, remains always fit for motion.

CHAP.

CHAP. IV.

Some further observations on the coagulable lymph, and on the sudden changes produced upon it.

IF the reader has been persuaded of the common opinion, that the disposition of the blood to coagulate is increased in inflammatory disorders, it may perhaps appear to him, as it formerly did to me, a very extraordinary circumstance that the contrary should be true; and likewise that the blood should in reality be the more disposed to concrete, in proportion as the body is weakened, or as the action of the blood-vessels is diminished. And as we are naturally tenacious of old opinions, and unwilling to adopt new ones

till

till fully proved, he may fufpect that there has been fome fallacy in thefe experiments. And indeed I muft acknowledge, that there is, in appearance, one ftrong argument againft my general conclufion, which is, that it has not only been remarked, that the firft cup has a cruft, whilft the laft has none; but likewife, that the fecond, or the third cup, alone fhall have a cruft, whilft the preceding ones are without it. Now this, I fay, feems contradictory to what I have advanced, concerning the difpofition of the blood to coagulate being increafed in proportion as the body is weakened; for here in proportion as the blood is evacuated, its difpofition to coagulate is leffened; fince it was more fizy in the fecond, or third cup, than in the firft. But, in anfwer to this objection, I muft remark, that thefe cafes very feldom occur; and that in general the firft cups are more fizy, and are the lateft in jellying; and when the contrary takes place,

place, or when the second or third cup is more sizy than the preceding, I am persuaded, that upon a careful examination, instead of weakening, they will be found to strengthen my inference; as will appear probable by the following case, which has occurred since these experiments were published in the Philosophical Transactions.

EXPERIMENT XXVII.

ON the 13th of June, I visited a young man, twenty-two years old, of an athletic habit, who complained of a violent pain in his head and back, with a full strong pulse; but as he was then in a profuse sweat, which had been preceded by a shivering, it was not thought proper to bleed him, and the rather, as we were informed, that he had had a similar paroxysm two days before. But next day, finding that his fever had not left him with the sweat, and that he still had

a pain

a pain in his head and back, and that his pulse, though not now full and strong, yet was quicker than natural, it was then judged necessary to take away some blood. Upon opening the vein, the blood flowed very slowly, and indeed merely trickled down his arm. Imagining that the bandage might be too tight, I slackened it, but still the motion of the blood was not accelerated. I then asked him whether he had not been afraid of the bleeding, and he told me he had; and on feeling his pulse in the other arm, I found it very low. I therefore desired him to move the muscles of his hand, which he did; but nevertheless so slowly did the blood run, that it was four minutes before I got an ounce and an half into a cup. I then stopt the orifice till another cup was brought, into which the blood ran in a full stream, to the quantity of three ounces, and that in two minutes, although the orifice was rather small, so much was its velocity now increased.

creaſed. Into the third cup, which likewiſe held three ounces, the blood ran ſtill faſter, as it was filled in leſs than two minutes. By this time the patient beginning to be faint, I ſtopt the bleeding till he could lie down on the floor, and then about three drachms more of blood were received into a fourth cup: this came away very ſlowly, and the bleeding ſtopt of itſelf. He drank a glaſs of water, and did not faint, and he appeared afterwards to be much relieved by the evacuation. Upon this blood I made the following remarks:

That which was taken away laſt was firſt coagulated, and completely too, by the time I had tied up his arm, which was in three minutes from the blood's firſt running into the cup.

The blood which was received into the firſt cup coagulated next, and as I obſerved by my watch, in twelve minutes from its being ſet down on the table.

That

THAT which was received into the second cup was the third in order as to coagulation, and was confiderably later in jellying than the firſt; for in fifteen minutes it was not thoroughly coagulated; nay, even in twenty-two minutes a ſmall part of it was ſtill fluid. It was remarkable, that none of theſe three had any ſize.

BUT the blood in the third cup differed confiderably from that in the others; for in five minutes it began to appear tranſparent on its ſurface, an indication of a future ſize, and it was later in coagulating than that in the other cups; for even at the end of twenty-ſix minutes a great part of the coagulable lymph was ſtill fluid, as appeared on removing the pellicle that covered it; but in thirty-five minutes it was completely jellied. The ſize in this blood was very thick and tough.

Now this cafe, when carefully examined, inftead of being an objection to my conclufions, will, I prefume, be thought a ftrong confirmation of them.

For, in the firft place, as the blood in the third cup alone had a cruft, and was much later in jellying than the reft, it ftrengthens my inference, that the difpofition of the blood to coagulate is leffened in thofe cafes where the inflammatory cruft or fize appears. And as the blood ran more rapidly into this cup, it fhewed that the heart and blood-veffels had begun to act with greater force, and therefore confirmed the opinion, that in proportion as thefe act more ftrongly, the difpofition of the lymph to coagulate is diminifhed. The fame opinion is likewife fupported by obferving what happened to the blood in the firft cup, which coagulated fooner than that in the third, owing to the veffels then acting more weakly,

weakly, as was evident from the blood's trickling down the arm, and from the lowness of the pulse *.

2*dly*, It may be observed, that the great difficulty in admitting the conclusion made in the former part of these

---

* In like manner may be explained another variety in the appearance of the size, namely where it is found in the first and last cups, but not in the second or third: this I suspect seldom happens, but when it does, it may perhaps be found, on examination, that the vessels were acting more weakly whilst the second or third cups were filled. For, so easily does this size appear to be removed, or formed, that I suspect it may sometimes happen, that when the blood is taken away, in a full stream, from a large orifice, the patient may be so suddenly weakened, and the properties of the blood may in consequence be so changed by the time the second cup is filled, that the size shall be removed: and yet afterwards the vessels may recover their former tone, so that the third or fourth cup may acquire a size again. Nay, I suspect that this appearance may even be affected by the passions, particularly from observing that the patient abovementioned, as well as others whose blood at first trickled slowly down their arms, had been much afraid of the lancet.

sheets

sheets (viz. that the want of size in the last cup is occasioned by an alteration in the blood-vessels) was to conceive how these vessels could possibly alter the properties of the lymph so suddenly, as in the time between receiving the blood into the first cup, and into the last. But this case confirms that inference, by shewing the fact in a clearer point of view; for even here, where the appearance of the size was reversed, it was found that the blood which had a crust or size was latest in coagulating, and that it was this blood which was taken out of the vessels when they acted most strongly, as was proved by the rapidity with which it flowed into the cup.

3*dly*, Since the times in which the blood jellied in these cups were so very different (the first coagulating in twelve minutes, the second in about twenty-two, the third in thirty-five, and the fourth in less than three minutes, notwithstanding these thefe cups were filled in lefs than two minutes after one another), it fhews, I fay, how foon that ftate of the blood-veffels on which the fize depends, can be removed and affumed, and therefore leads us to conclude, that although this fize is in general a fign of an inflammatory diforder, or a ftrong action of the veffels, yet there may be feveral circumftances to be taken into the account, before we can judge from its prefence, or abfence, whether or no venefection fhould be repeated: and it likewife fhews clearly, that it would be improper to determine, from the prefence of this alone, when bleeding is neceffary; and yet there have been not a few who have inclined to make fuch a conclufion, from their confidering this cruft or fize as fo very morbid an appearance.

4*thly*, As the blood in the third cup was fo late as thirty-five minutes in coagulating, and was fizy, whilft that in the fourth

fourth was not so, and jellied in less than three minutes, although it had been taken from the vessels only two minutes after the other, but at the time the patient had become faint; it shews how much faintness and languor increase the viscidity of the blood, and likewise its disposition to coagulate, since in two minutes they produced such a change as to remove the size, and to reduce the time of coagulation from thirty-five to three minutes. It therefore shews clearly how much languor and faintness should be encouraged in hemorrhages, and how carefully we should avoi giving any thing that can stimulate, or rouse the patient; that the medicines likely to be of service are nitre and the acids; or such as cool the body, or have the property of diminishing the force of the circulation, or of increasing that languor or faintness *;

that

---

* It has been objected here, that nitre would seem improper for this purpose, because in experiments

that all agitation of mind fhould, as much as poffible, be prevented, left it increafe the circulation: that all mufcular motion fhould be avoided for the fame reafon: for that an exertion of the patient's ftrength can leffen the difpofition of the blood to coagulate, I am perfuaded from fome of the abovementioned cafes, and likewife from what I have obferved in dying fheep, where the ftruggles of the expiring animal feemed in fome inftances,

ments mentioned before (p. 12.), it was found to prevent the coagulation of the blood, out of the body; but this objection is removed, by confidering, that, in order to prevent coagulation, the nitre muft be ufed at leaft in the proportion of two fcruples to every two ounces of blood. But, when we exhibit it internally, we feldom give more than a fcruple every two hours, which can have no effect in attenuating the whole mafs of blood, nor in preventing coagulation; efpecially as we have reafon to believe its properties are changed, before it paffes the digeftive organs. Its good effects in hemorrhages, therefore, are probably owing to its action upon the ftomach. For proofs of its utility, fee *Medical Obfervations* and *Inquiries*, vol. IV. art. xvi.

when

when violent, to alter the properties of the lymph.

We have endeavoured to explain the appearance of the inflammatory cruſt or ſize, from the red globules having ſubſided from the ſurface of ſuch blood before it coagulated: this we obſerved was partly owing to the lymph's being later in coagulating in thoſe caſes, but principally to its being thinned. But we may now add, that although the attenuation of the lymph, and its leſſened tendency to coagulate, are connected in moſt of thoſe caſes, yet they do not always go together; for the lymph may have its diſpoſition to coagulate leſſened without being thinned; which was evident in the preceding caſe, on comparing the blood in the ſecond with that in the third cup; for the blood in the ſecond cup had no ſize, notwithſtanding it remained fluid at leaſt ten minutes after the ſize had begun to appear in the third: this I attribute to the blood in the third

being

being more attenuated, and thereby more readily allowing the globules to subside.

That the blood may have its disposition to coagulate leſſened, without being attenuated, is likewiſe probable from the following caſes.

EXPERIMENT XXVIII.

In the month of January I bled a man, who complained of a pain in his head, attended with giddineſs and ſhivering, a pain and ſickneſs at his ſtomach, and with a full and quick pulſe; the blood was found to remain fluid for ten minutes, and then jellied, but no ſize appeared.

EXPERIMENT XXIX.

In another perſon, who was bled merely for a drowſineſs, and becauſe he was accuſtomed to that evacuation in the Spring,

Spring, I found the blood remain seven minutes without coagulating, and yet it was without any size.

Now, since in these cases the blood remained so long fluid, and yet the red particles did not subside, or no size appeared, I should conclude, that only the disposition of the lymph to coagulate was lessened, without its being thinned. And from the last case we may likewise conclude, that although the times, at which the blood taken from persons in health begins to coagulate, be allowed to be about three minutes and an half, as I have found from repeated observations, yet there may be some variety in this respect; for a *plethora* and other circumstances may make it later in coagulating in some cases, even where the patient is otherwise in perfect health *.

* This inference is confirmed by a case mentioned below, Experiment 31.

We have obferved before, that the fize is fometimes very firm, and at other times fpongy and cellular; thefe differences in its denfity are, I fufpect, in proportion to the degree of attenuation and leffened difpofition of the blood to coagulate; for as the coagulation begins on the furface, and forms there a film which attracts the reft of the lymph, the more that lymph is attenuated, and the flower it coagulates, the more will the film be able to feparate it from the red globules, and from the ferum: thence perhaps it is, that when the blood, befides being very thin, likewife jellies flowly, we fometimes fee almoft the whole coagulable lymph collected at the top, forming a firm cruft, which being free from the ferum, as well as from the globules, contracts the furface into a hollow form. But when the blood has its difpofition to coagulate lefs diminifhed in proportion to the attenuation, then, although the globules fubfide from the furface, yet the whole

whole of the lymph jellies fo foon after the coagulation begins, that there is not time for its being feparated from the ferum, of which it therefore contains a confiderable quantity, and is of courfe more fpongy and cellular.

In proportion to the thicknefs and denfity of the fize, the bottom of the cake is of a loofer texture; but this loofenefs of texture is not owing to putrefaction, as has been fufpected, but merely to the lymph's being collected at the top, and therefore leaving the bottom of the *craffamentum.*

Notwithstanding bleeding does in general weaken the action of the veffels, increafe the difpofition of the blood to coagulate, and even thicken the lymph; yet it may happen, that, in the ordinary quantity in which blood is taken away, none of thefe effects fhall be produced; of this the following cafe feems to be an inftance.

EXPE-

EXPERIMENT XXX.

A woman in the seventh month of her pregnancy was bled for a violent pain in her side, attended with a cough; the quantity taken away was eight ounces, which was received into four cups; and as the orifice was small, about ten minutes were spent in the bleeding. On attending to the different cups, I could observe no difference in the periods at which the coagulation commenced, and finished in each, allowance being made for the time the blood began to run into each. In every one of these cups the blood was completely jellied in about twenty minutes, and each had a crust or size nearly of the same thickness. So that the bleeding seemed not to have produced any change in the strength of the patient's vessels, nor was her pain sensibly abated by it. She was therefore desired to live low, to confine herself to a vegea vegetable diet, and to take a scruple of nitre every three hours in a draught of the *decoctum pectorale*; and if her pain and cough were not abated in a day or two, she was directed to repeat the bleeding. As close attendance was not required, I did not visit her till four days after, and then she had got free of her complaints, notwithstanding her blood had been apparently so little changed in the time of the evacuation.

In this case the bleeding seemed neither to have thickened the lymph, nor increased its disposition to coagulate, nor weakened the action of the vessels; but that it generally produces these effects, cannot, I think, be doubted, from our having observed it in so many instances. Perhaps the dread of the operation might here have made the coagulation of the blood in the first cup approach nearer to that in the last; or perhaps the smallness of the orifice prevented there being so manifest

manifest a change produced by the evacuation, from its giving time to the blood-vessels to adapt themselves more equally to the quantity they contained, by which means she was not weakened by the loss of blood.

It has been observed by Sydenham and others, that it sometimes happens, even in inflammatory disorders, when the blood trickles down the arm, instead of running in a full stream, it does not acquire a crust or size *. May not this

be

---

* It may be necessary to observe, that it is not in every case where the blood trickles down the arm that it is without a size; on the contrary, it sometimes happens, that even in such circumstances it has a very dense one; an instance of which may be seen below, in Experiment xxxi. In those cases the trickling down the arm may perhaps be owing to some circumstance in the orifice preventing its flowing in a full stream, or to a difference in the tightness of the ligature, rather than to a weak action of the vessels. Or, although the size be occasioned by a strong (or some particular mode of) action be explained from what is obferved in the cafe related in Experiment xxvii? that is, in fuch inftances the veffels, either from a febrile, or from fome other oppreffion, act more weakly than they do in the ordinary cafes of inflammation; by which means the lymph is not fufficiently attenuated to allow the red globules to fubfide before the coagulation begins, and therefore the fize does not appear, as in other cafes of inflammation where there is no fuch oppreffion.

As air is found to coagulate the blood and cold to thicken it, an objection has thence been made to the conclufions from fome of the preceding experiments; and

---

action in the veffels, and therefore is removed by weakening them, yet it may not always be removed immediately on their being weakened. For it may happen, that in fome cafes the lymph may not be fo fufceptible of changes as in others; or when it has been very much attenuated it may not again be thickened immediately, on the veffels acting weakly.

it has been fuppofed, that the changes in the properties of the blood, that happen during the time of bleeding (which I have attributed to a difference in the action of the veffels) might poffibly be owing merely to a difference in the expofition to the air, or to cold. For inftance, fince the blood that trickles down the arm feems to be more cooled than that which flows in a full ftream, it has thence been fuppofed, that its want of fize, in thofe cafes, might be owing to the expofition to the air, which made it coagulate fooner, and to the cooling which had thickened it, and thereby prevented its red particles from fubfiding fo that the fize fhould be formed. This objection is indeed plaufible, and to thofe who have not feen thefe experiments, might at firft feem fufficient to explain the appearance; but upon further examination it will not be found to do it fatisfactorily. Thus, for example, although it be true that air coagulates the blood, and likewife cools it, yet there

there are changes remarked in the preceding experiments that cannot be explained merely by a difference in the expofition to the air: Thus, in Experiment XXVII. the blood in the third cup was thirty-five minutes in being completely coagulated, whilft that in the fourth, although taken from the arm only two minutes later, yet coagulated in three minutes. Now no expofition to the air, nor to cold, from the blood's trickling down the arm, could produce fuch a change. Of this I am perfuaded from what I have obferved on comparing the blood received into a cup, with that which dropt on the plate which held the cup; for I have repeatedly feen on thofe occafions, that the blood on the plate, although it was fo much more cooled and fo much more expofed to the air, than that in the cup, yet inftead of coagulating proportionably fooner, was later in being coagulated. The following experiment fhews this clearly.

EXPE-

EXPERIMENT XXXI.

A YOUNG woman with a violent inflammation in her eyes, was bled on the 5th of March, early in the morning, before she had breakfasted, and whilst she was complaining of a sickness at her stomach; the blood followed the lancet in a stream, but immediately after it only trickled down the arm, and continued to do so during the whole of the evacuation. About eight ounces of blood were taken away into four vessels, viz. into two cups and two saucers, in the following manner: A plate holding both a cup and a saucer was held under the arm, and the blood was first received into the saucer, to the quantity of a spoonful, then as much more was received into the cup that stood by it; then again the blood was suffered to run into the saucer, and afterwards into the cup, and so alternately till there was about two ounces in each, when they were carefully set down on a window where the thermometer stood at 57°; the plate was placed by them, and contained about a spoonful of blood, which had missed the saucer in the beginning of the evacuation. Next, the second plate was brought, and some blood was received first into the cup and then into the saucer, in the same manner; and three portions of blood were suffered to drop at different times on the plate, each of them about the breadth of a shilling. Now, here, according to the reasoning in the objections made to some of the preceding experiments, the blood in the saucers having twice as much surface as that in the cups, ought to have coagulated in half the time; and that on the plates ought, from the largeness of the surface, to have coagulated in much less time; but just the contrary happened; for the blood in the cups was first completely coagulated, that in the saucers next, and that on the plates latest of all. But, as as the experiment seems curious, it may be proper to give a more particular detail of what was observed.

On looking at the first plate, at the end of seven minutes after it was set down on the window, the surface of the blood in the cup was considerably transparent, and a pellicle (that is the surface beginning to coagulate) was formed upon it; but no transparency was distinguishable on that in the saucer, nor could any pellicle be observed upon drawing a pin through it, or through that which had dropt upon the plate. At the end of fifteen minutes the blood in the cup and in the saucer were pretty much coagulated, or had a thick pellicle, whilst none could yet be observed on that upon the plate. At the end of fifty-five minutes that in the cup was just beginning to part with its serum, whilst the blood in the saucer was not yet completely coagulated; for on inclining

ing it to one side, a part of the blood appeared fluid under the pellicle. That on the plate was now coagulated. They were all three sizy; and the blood in the saucer had a size which seemed to be as firm, and in as large a quantity, nearly, as that in the cup; and the size upon the blood in the plate was thick enough to be easily distinguished.

In the second plate, at the end of seven minutes after being set on the window, both the blood in the cup and in the saucer were beginning to coagulate; and had a pellicle of a considerable thickness, and were both sizy: but no pellicle appeared on any of the three portions that had dropt on the plate. At the end of fifteen minutes, that in the cup was firmly jellied, that in the saucer not quite so much, and one of the spots on the plate was but just beginning to coagulate at its edge. At the end of twenty-five minutes, the two last spots were

were still perfectly fluid, but in twenty-eight minutes they were beginning to coagulate; whilst the blood in the cup was now parting with its serum. At the end of fifty minutes a considerable quantity of serum had separated in the cup, and the separation was just beginning in the saucer.

This experiment was repeated on another person's blood two days after, in the presence of Mr. Field and Mr. Hendy, two studious gentlemen, at that time living at the Middlesex Hospital, and the appearances were exactly similar; and it was evident to them that the blood in the saucers was later in coagulating than that in the cups, and that on the plates (one of which was of pewter) was considerably later in jellying than that in the cups or in the saucers.

These experiments therefore shew clearly, that the differences in the periods

riods of coagulation, and in the appearance of fize upon the blood received into the different cups in bleeding, cannot be accounted for from a difference in the expofition to air; for here blood more expofed to the air than that is which trickles down the arm, is found equally fizy, and to be even later in coagulating than blood lefs expofed.

As we have here obferved a new circumftance that appears remarkable, and which at firft fight feems not reconcilable to fome of our conclufions, it may therefore be neceffary to examine it farther, or to enquire, If air be a coagulant of the blood (as we have endeavoured to prove in the beginning of this Effay) how comes it that in this experiment the blood was not coagulated proportionably to its expofition to the air? This, I think, may be explained from confidering another fact that was mentioned in the preceding pages, *viz.* that cold

cold leffens the blood's tendency to coagulation. The blood, therefore, in the faucer, although it was more expofed to the air, yet being more cooled than that in the cup, was, for that reafon, later in coagulating; and that on the plate, which was moft expofed to the air, being moft cooled, was therefore lateft of all.

But we may add, that although it be evident from this experiment, that the difference in the expofition to air, or to cold, is not fufficient to explain the changes which we fee produced upon the blood, in fo little time as in the filling of a fmall cup, efpecially when thofe changes are fo great as what are mentioned in Experiment xxvii. (where the blood in one cup was thirty-five minutes in coagulating, and had a very thick fize, whilft that taken away foon after, coagulated in three minutes, and was without a fize) yet, I think, that the effects of air, and of cold, are confiderable enough to deferve to

be

be taken into the account in some cases, where the changes on the blood are not so great. For as cold thickens the blood, it is probable that in some cases where the lymph is but little attenuated, and where therefore, in the ordinary manner of bleeding, there would have appeared but very little size, such blood, if more exposed to the air so as to be sooner cooled, may thereby have the small degree of attenuation counteracted, or removed, and the red particles may be prevented from subsiding. So that although in this experiment, where the size was thick, it appeared equally in the saucer as in the cup, and even appeared on the blood upon the plate; yet, if we repeat these experiments on a variety of subjects, it is probable, that we may sometimes find the saucer without a size, whilst the cup has one; for when the lymph is but little attenuated a slight cause may thicken it again; and its being a little more cooled in the saucer

and

and on the plate, may in some cases be sufficient to prevent the size from appearing.

It has likewise been suggested, that possibly there might be some difference in the orifice, from which the blood flowed, to which its different appearances in the several cups might be owing. But there does not seem to be any foundation for this objection, and it may, I think, be removed by a careful examination only of some of these experiments, particularly the 27th; for there the blood ran in a full stream both into the second cup and into the third, the orifice being apparently unaltered, and yet there was a great difference in the appearance of the blood: for that in the third cup had a thick size, but that in the second had none. So that there does not seem to be any circumstance attending these experiments that can explain the changes produced upon the blood

blood in the time of bleeding, excepting that to which I have attributed it, *viz.* a change in the ftrength of the blood-veffels, or in their mode of action; and every obfervation I have yet made confirms me in that opinion.

A very eminent phyfician ‡, after reading the firft edition of thefe fheets, informed me, that from a fuggeftion which he met with in Profeffor Simfon's Differtations, *viz.* " That a ligature on the " arm would produce a fize †," he had been for many years cautious how he took any indications from this appearance of the blood, and defired I would endeavour to determine, by experiment, whether the ligature's being a longer or fhorter time upon the arm, even in the

‡ Dr. Fothergill.

† Dr. Simfon's obfervation is, that if a tight ligature be made on the limb, and the vein opened three hours after, a fize will be produced.—*Vide De Re Medica, Differt.* iii. *Sect.* 38. p. 112.

ordinary

ordinary way of bleeding, might not influence this appearance of the blood. And accordingly in the prefence of Dr. Drummond, Mr. Field, Mr. Hendy, and Mr. Cockfon, I made the following experiment:

EXPERIMENT XXXII.

ON the 9th of October, I tied up both the arms of a healthy young man with a degree of tightnefs fufficient to make the veins fwell and become turgid, whilft the pulfe remained free; a vein in one arm was opened immediately after making the ligature, and an ounce of blood was received into a cup. I chofe to take away fo fmall a quantity that there might be the lefs probability of producing any change upon the blood by weakening the body. The ligature was left upon the other arm for an hour, which made the veins very turgid, and likewife made the perfon complain of a

stiffnefs

stiffness in his fore-arm; the artery in his wrist being felt all the time, but less distinctly than in the other arm which was without a ligature. At the end of an hour this vein was opened, the orifice was large, and an ounce of blood was taken away. Upon attending carefully to each cup, it did appear that the ligature had produced some change; for in the first place, the blood which had been so long detained in the arm by the ligature was darker coloured, or blackish, whilst that from the other arm was more florid, even at its first running from the vein. 2dly, The blood that had been so long in the arm was rather later in being coagulated; for it did not begin to part with its *serum* till at the end of thirty-seven minutes after puncturing the vein; whilst in the other the coagulation was completed, and the *serum* was beginning to ooze out in thirty minutes. 3dly, The blood which was first taken away was without a size, whilst that which had been so long in the other arm had a small spot

spot about the breadth of a silver penny, and did not cover a twentieth part of the surface.

From this experiment, therefore, it would seem, that a ligature long continued may produce a size, agreeably to Dr. Simson's observation, but then it will probably be only in small quantity.

The same learned gentleman, on being informed of the result of this experiment, ingeniously suggested, that the quantity of blood which I had taken away might perhaps be too little to make the experiment decisive; for, as only one ounce of blood had been taken from each arm, all that blood might be supposed to have been contained in the veins themselves; and as it was more probable that the disposition to size took place in the arteries, a larger quantity should be taken away, in order to judge whether the blood in the arteries had not been changed by the veins of the arm being so long compressed. The experiment was, therefore, repeated upon the same person, on the 7th of March, Mr. Field and Mr. Hendy assisting me as before, and we observed as follows:

EXPERIMENT XXXIII.

THE blood from the arm first opened was in quantity about ten drachms, had no size, but was late in being completely coagulated. The pellicle first appeared on its surface six minutes after opening the vein, and at the end of fifteen minutes a considerable quantity of the blood was still fluid; but in thirty-four minutes it was completely coagulated. The *serum* did not begin to ooze out till at the end of fifty minutes.

AFTER the other arm had been tied up an hour, the vein was opened, and about ten drachms of blood were received into the first cup; as much more into a second;

second; and about an ounce and an half upon a pewter plate; and about two ounces and an half into a third cup (in all six ounces and an half) and it was observed, that the blood in the third cup and that in the second coagulated in about twenty minutes, and the *serum* began to ooze out in twenty-five minutes after opening the vein; the blood on the plate was later in coagulating, and none of these portions of blood had the least appearance of size. But the blood in the first cup was considerably the latest in jellying; for at the end of twenty-five minutes a large quantity was still fluid under the pellicle, and even at the end of fifty minutes, the coagulation was incomplete. This blood seemed to have rather more size than that in the former experiment, but it was not collected at one part, but was diffused over the surface, and appeared in spots not bigger than pins heads.

The result therefore of this experiment was similar to the preceding. The ligature long continued produced a size, but in small quantity; and therefore it does not appear probable, that, when the ligature has been only a few minutes on the arm (as is the case in the ordinary manner of bleeding) it can deserve to be taken into the account, when we judge of the disease, or of the indications of cure from the appearance of the size: especially when it is considered, that the blood on which these experiments were made was very favourably circumstanced for Dr. Simson's opinion; that is, it was from a patient who seemed to be plethoric, by the blood which was first taken away not jellying in less than thirty, and thirty-four minutes, which is later than ordinary, as appears from what is observed above, page the 40th; where we found, that the blood of people in health commonly jellies in seven or eight minutes.

As

As many of these experiments shew how easily the disposition of the lymph to coagulate can be altered, even by slight changes, as it would seem, in the state of the blood-vessels, they help us to explain how it should happen, that the blood, in some diseases, is found without this property of jellying; an instance of which is mentioned by Mons. de Senac*; another was observed by the learned professor Cullen; and a third I saw lately by the favour of Dr. Huck and the physicians of the British lying-in hospital, who, having bled a woman in a fever that came on soon after delivery, found her blood did not coagulate on being exposed to the air, but appeared like a mixture of the red globules and *serum* only, the globules having subsided to the bottom in the form of a powder. She died three days after, and, upon opening her, we found the blood had coagulated in her vessels after death, and

---

* Traité du Cœur, Tom. 2. p. 129.

that a tough white *polypus* was formed in each auricle of the heart, one of which I have now by me. I examined the blood taken away before she died, and found, on exposing it properly to heat, that it did not coagulate sooner than *serum* commonly does, nor under 160°; so that it is probable, that, at the time she was bled, her blood either was without the coagulable lymph, or its properties were changed.

After a blow or contusion, the blood now and then bursts from the vessels into the cellular membrane, sometimes forming an *ecchymosis*, and sometimes a tumor, and it is a question with some, whether such blood coagulates or not; but that it coagulates in most of these cases, is proved by opening such tumors. Yet it has likewise been observed, that now and then these tumors have been attended with a fluctuation, and that, after some time, their contents have been absorbed, forbed, and it has also been found, that, upon opening some of them, even several weeks after the accident, the blood was fluid. In such cases the blood had probably undergone a change similar to what was observed to take place in some of the preceding experiments, that is, the lymph had been deprived of its property of coagulating, in passing from the blood-vessels into the tumor: a circumstance, which, how remarkable soever it may appear, agrees with what we have above observed of the lymph, whose properties seemed to vary with the state of the blood-vessels.

CHAP.

CHAP. V.

Containing a recapitulation of the principal facts and conclusions that are mentioned in the preceding pages.

THE separation of the blood into *crassamentum* and *serum*, in a given time, appears to be in proportion as the heat in which it stands is nearer to that of the human body. The heat in which the blood is kept should therefore be attended to, when we draw inferences from the proportions of these two parts.

THE florid colour of the surface of the *crassamentum* seems to be owing to the air. The venous blood, in passing through

through the lungs, has a similar change produced upon it, or becomes more florid by the time it gets into the arteries; and this florid colour is again lost in passing from the small arteries into the veins, especially if the person be in health. But sometimes in diseases it does not undergo this change, but comes florid out of the veins.

Neutral salts, on being mixed with the blood out of the body, make it more florid: they likewise, if used in great quantity, prevent its coagulation when exposed to the air, and some of them allow the lymph to be precipitated, or to jelly on being diluted with water. But we are not to conclude from thence, that they would produce the same effects when used as medicines; for then they are only given in small quantities, and may have their properties changed by the digestive organs, before they are mixed with the blood. And as they are most of them cooling and sedative, they produce

produce naufea and languor of the ftomach, and leffen the force of the vafcular fyftem. And as the properties of the blood feem to depend on the action of the veffels, thefe falts may thus, by affecting the veffels, produce changes on the blood very different from what might have been fufpected, from obferving what takes place when a large quantity of them is mixed with the blood out of the body.

The blood is not coagulated (I do not mean thickened, for it is indeed thickened) by cold, but on the contrary, has its difpofition to coagulate leffened, and even entirely taken off, if the expofition to cold be long continued. When therefore the blood in the bafon jellies, it is not the cold that produces this effect, nor is it the want of motion; for although the blood by being at reft will jelly at laft, yet it will not do it in the time the coagulation takes place in the bafon;

bafon; which in the blood of healthy perfons is in feven or eight minutes after being received from the vein. The coagulation of the blood in the bafon is therefore probably owing to the air.

When the blood is at reft in the body, it will at laft coagulate merely for want of motion; but this coagulation takes place with different appearances from that of the blood in a bafon, for it begins in ten or fifteen minutes, and is not completed in three or four hours; whilft the blood of the fame animal, taken out of the veins, and expofed to air, will begin to jelly in three or four minutes, and will be completely jellied in feven or eight.

The effect that air has upon the blood is not immediate on its application, but takes place fooner or later, in diffcrent circumftances of health; in fome cafes the blood is coagulated in a few feconds after it has been expofed to the air, in others not in lefs than an hour and an half, or perhaps more, as appears from Experiment XIV.

The inflammatory cruft or fize is not a new-formed fubftance, but is merely the coagulable lymph feparated from the reft of the blood. This feparation feems to be occafioned by the lymph's being attenuated, by which means the red particles foon fettle to the bottom, and leave the furface of the blood tranfparent; and this tranfparent part being a mixture of the coagulable lymph and the ferum, the former coagulates on its furface where, in contact with the air, and the difpofition to coagulate being likewife diminifhed, the blood remains a long time fluid, and thereby gives time for the pellicle formed on its furface to attract the reft of the lymph, and to collect it at the top, leaving the bottom of the cake proportionably fofter. The fize therefore

fore is thicker, and denfer, in proportion as the lymph is thinned, and is lefs difpofed to concrete. But it is not a certain fign of inflammation, and does not appear to be the caufe of that difeafe, but only its effect.

But the moft remarkable conclufion that thefe experiments have led to, is, that the properties of the blood depend on the ftate of the blood-veffels, or that they have a plaftic power over it, fo as to be able to change its properties in a very fhort time. The novelty of this opinion, and the difficulty of conceiving how the veffels fhould have fo remarkable a power, has made fome object to the conclufion, who had not well confidered all the facts from which it was deduced. I fhall here therefore fum up thefe facts.

That the blood-veffels, by their ftronger or weaker action, can change the properties of the lymph, even in the fhort short time spent in filling the different cups in bleeding, is first inferred from Experiment XIX. where the blood in the first cup had a size, whilst that in the others had none. Now as this want of size in the last cups was owing to the lymph's having by this time become thicker, and to its being more disposed to coagulate, and no other circumstance being observed that could account for this alteration (for the difference in the exposition to cold, or to air, even then, appeared inadequate to the effect, and are now proved to be so by Experiment XXXI.) I concluded, that it was entirely owing to an alteration in the strength with which the vessels acted upon the blood. And to this opinion I was led by the well-known fact, that bleeding weakens the body; and as bleeding, which weakens the body, had here removed the size, by thickening the lymph, and by disposing it more to coagulate, I thence inferred

that

that it was by its weakening the body, or the action of the veffels, that it had produced thofe changes on the lymph. The confideration of what takes place in inflammations confirmed me in this opinion; for in them the blood-veffels are known to act more ftrongly, and it is proved above, that the lymph is proportionably thinned, and has its difpofition to coagulate proportionably diminifhed.

AND this opinion was further ftrengthened, by obferving what occurs when an animal is bled to death, or when the veffels are acting with the loweft degree of ftrength; for here it was found, that in proportion as the ftrength of the animal was reduced, the blood was more and more difpofed to coagulate.

AND having thus obferved the connection between the alterations in the lymph, and changes in the ftrength of the

the blood-veſſels, in theſe caſes, I next inferred, there might be the ſame correſpondence, even in others where the changes in the properties of the lymph are more ſudden, as in Experiment XXVII. where there was no ſize in the firſt cup, but a thick one in the third; and even this caſe, when carefully examined, confirmed the inference; for the blood-veſſels were found to be acting with different degrees of ſtrength, at the very time that the lymph was found to have different properties. And the only difficulty that remained here, was to explain how it ſhould happen, that the firſt cup, contrary to what in general takes place, ſhould have its blood coagulated in leſs time than the ſecond or third; and this I concluded was owing to ſome febrile cauſe, affecting or oppreſſing the patient; in which concluſion I am confirmed by the fact admitted amongſt phyſicians, that the pulſe is frequently oppreſſed in inflammatory diſorders,

the Properties of the Blood.

diforders, and rifes in ftrength in proportion as blood is taken away.

From this conclufion I went further, and conjectured that even temporary exertions of ftrength in the blood-veffels might alter the properties of the lymph; to which opinion I was led by having obferved the blood fizy in the cafe mentioned in Experiment XX. where great weaknefs foon followed the evacuation; and likewife from having obferved, that the ftruggles of dying fheep feemed to alter the lymph.

So that upon the whole, the opinion agrees with all the appearances, and is fupported by all the differences in ftrength that occur in the various deviations from the ftandard of health: for when the veffels act more ftrongly than they do in health, the lymph is proportionably more thinned, and is lefs difpofed to concrete; and when the veffels act more weakly than in health, then the lymph is proportionably thickened, and is more ready to concrete. Is it not therefore probable, that the differences we obferve in the thickness or thinness of the lymph, and in its being more or less difpofed to coagulate, are owing to thefe differences in the ftrength of the blood-veffels? For fuch alterations in the ftrength of the blood-veffels are always connected with thofe of the lymph, and we can obferve no other circumftance connected with thofe changes of the lymph that can at all explain them.

AND although it muft be admitted, that it is very difficult to conceive how the blood-veffels fhould do this, yet I fhould hope that ingenious men would not merely on that account reject my conclufion; but would confider, that as it is deduced from a number of experiments, as it agrees with all the appearances, and as it leads to an explanation

planation of many of them which we cannot otherwife account for; it may be well founded although it be difficult to be conceived. For there may be powers in the animal œconomy that are not yet dreamt of in our philofophy.

This obfervation of the properties of the blood depending on the ftate of the veffels, befides explaining many morbid appearances, leads to a further application; for we may thence be led to advance more rational and more certain rules for the treatment of hemorrhages. For as hemorrhages feem to be ftopped, partly by a contraction of the bleeding orifices, and partly by the coagulation of the blood, and as the difpofition of the blood to coagulate is increafed by weakening the body, and likewife the contraction of the bleeding orifices is promoted by the fame means, it is therefore evident, that the medicines to be ufed, fhould be fuch as cool the body, and leſſen the force of the circulation; and experience teaches us, that ſuch are the moſt efficacious.

It likewiſe ſhews, that all agitation of mind, and all bodily motion ſhould as much as poſſible be prevented; becauſe they increaſe the force of the circulation, and are thence unfavourable to the ſtopping of the hemorrhage. But that languor and faintneſs being favourable to the coagulation of the blood, and to the contraction of the bleeding orifices, ſhould not be counteracted by ſtimulating medicines, but on the contrary ſhould be encouraged. And as evacuations weaken the body more when they are ſudden, we ſee a reaſon why blood-letting ſhould be adviſable in hemorrhages, and why a large orifice ſhould be preferable to a ſmall one, when we want to produce that languor or faintneſs, or that weak action of the veſſels, ſo uſeful for the ſtopping of the hemorrhage.

Before

Before we conclude, it may be added, that the practice here propoſed is far from being new or uncommon; but as ſome have recommended oppoſite methods, and both parties have appealed to experience, where authority ſo equally balances authority, the young practitioner muſt be at a loſs which to follow, and for want of principles to direct his choice, may frequently adopt the worſt practice: witneſs the uſe of port-wine, and other ſtimulating aſtringents, which is ſo very common in moſt parts of England.

CHAP. VI.

Of the Serum *of the blood, and particularly of the milk-like* Serum.

THE *serum*, when separated from the *crassamentum*, by letting the blood rest in the bason into which it is received, is a fluid, apparently homogeneous and transparent; of a yellowish colour, saltish to the taste, in consistence thicker than water, and its specific gravity, according to Dr. Jurin, is to water as 1030 to 1000 [*].

[*] Philos. Transf. No. 361.

WHEN chemically examined the *serum* is found to confist of a mucilaginous fubftance, which is diffolved in a water that contains a fmall quantity of neutral falts. The mucilaginous fubftance of the *serum* agrees with the coagulable lymph of the blood in being fixed or coagulated by heat; but the degree neceffary for the coagulation of the *serum* is greater than that which is neceffary for fixing the lymph; for the lymph is coagulated by a heat between 114 and $120\frac{1}{2}$ degrees of Fahrenheit's thermometer; (fee Experiment IX.) whilft the *serum* requires 160° to coagulate it: (fee Experiment X.)

BUT the degree of heat neceffary for their coagulation, is not the only circumftance in which the lymph differs from the mucilage of the *serum*; they differ more remarkably in the former coagulating when expofed to air, whilft the *serum* undergoes no fuch change.

When the *serum* is coagulated by heat, a watery fluid can be preffed out of it; and this fluid the learned *M. de Senac* diftinguifhes by the name of Serofity.

This ferofity contains the neutral falts of the blood, and although it has been expofed to the heat of boiling water, yet it ftill contains a part of the mucilaginous fubftance, which is combined with the water in fuch a manner as not to be coagulated by the heat of boiling water, till a part of the water is evaporated by boiling; and then it coagulates and appears not very much unlike the mucus fpit up from the *afpera arteria* in a morning.

When the mucilaginous part of the *serum* has been coagulated by heat, it cannot again be diffolved in the Serofity, except by putrefaction or by the addition of fome chemical fubftance, and then it it differs from what it was before, particularly in its not being coagulable by heat.

But if the *ferum* be expofed to a lefs degree of heat than is required for its coagulation, for example, to that of 100°, it is gradually infpiffated into a brownifh, folid mafs, and this mafs can readily be diffolved again merely by the addition of water; and the *ferum* feems to poffefs the fame properties that it did before, particularly it is capable of being coagulated by heat. But care muft be taken not to add more water than it had loft, for if more be added than was evaporated, it alters its properties of coagulation.

In this circumftance of being infpiffated, and again rendered diffolvable in water, the *ferum* agrees with the white of an egg †, but differs from the coagu-

† See Newman's Chemiftry, Sect. IX.

lable

lable lymph, which even when mixed with a neutral falt, (viz. true Glauber's falt) cannot be infpiffated, and diffolved again without coagulation.

IF frefh *ferum* be diluted with an equal quantity of water, and then expofed to heat, it does not coagulate in that of 160°, as when undiluted, nor even in the heat of boiling water, as I have lately obferved, but it can now be boiled without immediately coagulating. And as the water evaporates, a pellicle is formed on the furface, which becomes thicker and thicker as the evaporation advances. This pellicle feems to be the mucilage coagulated, for it cannot again be diffolved in water like the infpiffated *ferum*.

MILK, when boiled, has its mucilage or coagulable part feparated in like manner, in the form of a pellicle, in proportion as the evaporation takes place.

And both milk and serum, whether diluted or not, agree in being coagulated by rennet *, when exposed to heat. So that milk seems to be made of the mucilaginous part of the *serum*, or is a diluted *serum*, with the addition of an expressed oil, or with a saccharine substance instead of the neutral salts.

But although there is an analogy between milk and diluted *serum*, in the circumstance of coagulation, yet they differ in another; namely, diluted *serum* can by a moderate heat be inspissated without coagulating, or forming any pellicle on its surface, but if milk be exposed to the same heat, it is not inspissated so completely; for a pellicle is formed on its surface, in proportion as the evaporation takes place, and this pellicle seems to be as perfectly coagu-

* Which is an infusion of the 4th stomach or the *abomasus* of a calf.

lated as if the milk had been expofed to a boiling heat; for it will not diffolve again merely by adding water, as infpiffated *ferum* does.

SERUM, therefore, by being diluted, comes near to milk in the circumftance of its coagulation. But the coagulable lymph cannot, by any art yet difcovered, be made exactly to refemble *ferum*.

THE mixture with neutral falts makes it indeed fo far approach to the nature of *ferum*, as not to be coagulated by expofition to the air; but it does not alter fo confiderably its property of coagulating by heat, for the mixture with Glauber's falts (in Experiment VII.) coagulated at 125°; and, I believe, would coagulate in a heat of 123°, if long expofed to it; whilft the pure coagulable lymph is fixed between 114° and 120$\frac{1}{2}$, and the *ferum* not under 160 degrees of Fahrenheit's thermometer.

Of the white Serum.

ALTHOUGH the *serum* of human blood be naturally tranfparent, and a little yellowifh, yet it is frequently found to have the appearance of whey, and fometimes to have white ftreaks fwimming on its furface like a cream, and now and then to be as white as milk, whilft the *coagulum* is as red as ufual. In all thefe three cafes of whitenefs, I have examined it in a microfcope with a pretty large magnifier, and have found it to contain a number of very fmall globules, although naturally, when tranfparent, no globules can be obferved in it, notwithftanding what has been affirmed by fome authors. Thefe globules differ from the red particles (improperly called globules) in their fize, which is much fmaller; and likewife in their fhape, which is fpherical, whilft the red particles are flat. They agree more with the globules of milk.

I have

I have compared them with those of woman's milk, and have found, that in the milk the globules are of different sizes, some being three or four times as large as others, and the smallest little more than just visible, when viewed with a lens of $\frac{1}{23}$ of an inch focus, whilst those of the white *serum* are more regular, and are all of them about the size of the smallest globules of milk. Of this white *serum* I have met with the following instances in books. In Tulpius, one instance *, in Morgagni two †, in the Philosophical Transactions some instances ‡, in Sckenckius's Observations two cases are related from other authors ‖. I have likewise heard of the same appearance having been observed by the learned Sir John Pringle, Dr. Pitcairn,

---

* Tulp. Ob. l. 1. cap. 58.
† Morgagni, Ep. XLIX. Art. 22.
‡ Philosoph. Transact. N° 100, and 442.
‖ Sckenckij Obs. lib. 3.

Dr.

Dr. Hunter, Dr. Watson, Dr. Bromfield, Dr. Garthshore, and Dr. Fothergil of Northampton. And other instances have lately occurred to persons of my acquaintance, who have favoured me with a short account of them.

Mr. French, apothecary in St. Alban-Street, having informed me, that he had some blood bye him, taken from a woman the day before, whose serum was as white as milk; he favoured me with a small quantity of it for examination, and with it the following particulars of the case. " Mary Rider, about twenty-
" five years of age, of a fresh com-
" plexion, and lusty, has not had her
" menses for these seven months. She
" discharges blood sometimes by vomit-
" ing, and sometimes by stool; com-
" plains of a pain in her left side, and
" in her stomach: she has an inclination
" to eat, but when she tries, she soon
" after loathes her food. She com-
" plains of great lassitude and sleepiness;
" her

" her pulſe is ninety-five in a minute.
" She has been bled twelve times within
" theſe ſix months, and every time the
" *ſerum* was as white as milk."

Mr. Robertson, apothecary in Earl-Street, acquainted me, that " Mr. Her-
" bert, a publican, of about thirty-five
" years of age, and corpulent, had been
" ſubject to a bleeding at the noſe, to
" the piles, and to ſuch profuſe ſweats in
" the night, as to be frequently obliged
" to change his ſhirt in the morning
" before he got out of bed, but that, for
" ſome time paſt, his ſweats had ceaſed.
" That, on September the 23d, he was
" ſeized with a bleeding at his noſe,
" which had been preceded by a pain in
" his head for two or three days. That
" his bleeding continued till he had loſt
" about two pounds of blood, and then
" ſtopt; and that the ſerum of his blood
" was as white as milk. That at ten
" o'clock the ſame night, the hemorrhage
" returned, and he loſt a conſiderable
                              " quantity;

" quantity; neverthelefs, it was thought
" proper to take fixteen ounces of blood
" from his arm, during which evacua-
" tion he fainted, but his bleeding at the
" nofe ftopt. That the *ferum* of this laft
" blood was likewife very white. That
" on the 25th, in the morning, he again
" complained of a pain in his head, and
" about ten o'clock his nofe began to
" bleed again; but the *ferum* now ap-
" peared no whiter than whey. That
" he continued to lofe blood during moft
" part of the night, fo that it was fup-
" pofed he could not lofe lefs than two
" or three pounds, the ferum all this
" time being a little whitifh, but fo little,
" that the bottom of the veffel in which
" it ftood could now be feen through it.
" That his bleeding returned repeatedly
" till the third of October, when it
" entirely ftopt, the *ferum* having be-
" come more tranfparent towards the
" laft."

L   Mr.

Mr. Eustace, apothecary in Jermyn-Street, sent me a phial of white *serum* from one of his patients, by trade a butcher. " This man," he told me, " was tall, of a strong make, a hard " drinker, subject to puke every morn- " ing, took little food, sweated a good " deal, but did not waste in his flesh. " He was bled for a slight *asthma* to " which he was subject, and of which " he had always been relieved by bleed- " ing. In other respects he was in a " good state of health, so as to follow " his business without much inconve- " nience."

Besides these cases, my friend Mr. Lambert, surgeon at Newcastle upon Tyne, told me, " that he had a patient " some years ago with a violent rheu- " matic pain in his hip, whom he was " obliged to bleed thrice, and every time " his *serum* was as white as milk, but " the *coagulum* of its natural colour.
" This

" This gentleman," Mr. Lambert adds, " was a free liver, of a full make, but " rather mufcular than corpulent, and " remarkable for being a great walker."

When I firft faw this unufual colour of the *ferum,* I was inclined to adopt the opinion of thofe who have attempted to explain it by the patient's being bled foon after a meal, or before the chyle was converted into blood. But afterwards, on confidering the cafes above related, I found this could by no means be the caufe, as none of thefe patients had taken a fufficient quantity of food to occafion this appearance; on the contrary, moft of them had a bad appetite, and had taken remarkably little food, and were fubject to vomitings. I therefore concluded it was owing to fomething elfe, and what confirmed me in this opinion, was an obfervation I had repeatedly made in diffecting geefe, whofe *ferum* I had frequently feen white, whilft

their chyle was tranfparent; although they had been killed only three or four hours after eating. And as the whitenefs, in all the cafes that I examined, was owing to a quantity of fmall globules like thofe of milk (which are known to be oily) I concluded that thefe in the human *ferum*, when white, were oily likewife, and recollecting to have read fomewhere of an experiment by which butter had been got from fuch human *ferum*, I tried, by agitating fome of it a little diluted, to feparate its oil, or to churn it, but without fuccefs. I then infpiffated fome of it to drynefs, and compared it with the natural *ferum* of human blood prepared in the fame way, and found it lefs tenacious, and much more inflammable; and when thus dried, its oil oozed out fo much as to make the paper in which it was kept greafy. Another portion of this white *ferum* being kept fome days, putrefied, and when putrid, it jellied as milk does when become four; but it
differed

differed from milk, in being extremely fœtid.

'Now, as the white globules appear from thefe experiments to be of an oily nature, and as it is improbable, from thefe patients having taken little food, and from the tranfparency of the chyle in birds, that this whitenefs of the *ferum* fhould be owing to unaffimulated chyle, accumulated in the blood-veffels; we muft therefore believe it to be owing to fome other caufe. And as we know there is a confiderable quantity of oil laid up in the cellular fubftance of animals, which is occafionally re-abforbed, is it not moft probable that this curious appearance was, in the abovementioned cafes, owing to fuch a re-abforption? And as all thefe patients had fymptoms of a *plethora*, and were relieved either by fpontaneous hemorrhages, or by blood-letting, is it not probable, that, to whatever purpofe the oil is applied in the body after it is re-abforbed absorbed from the cellular membrane, in these patients it had been re-absorbed faster than it was applied, and by that means was accumulated in their blood-vessels? This conjecture seems to be confirmed, from considering that in most of these cases the people were inclined to corpulency, and that two of them laboured under a stoppage of a natural evacuation *.

Another conclusion which these observations lead us to, is, that since the chyle of the birds which I dissected was not white, but transparent, at whatever time after eating it was examined,

---

* Although it appears probable that the whiteness of the *serum* in the above-mentioned cases was not owing to the chyle, yet I would not conclude that the chyle does not in the human subject occasionally colour the *serum*. We frequently observe the *serum* of such people as are bled a few hours after a meal, a little turbid, like whey, which I believe may be owing to the chyle. But if the milk-like serum was occasioned by a full meal, it is likely we should oftener see it than we do.

it

it follows, that the fat (in these animals at least) is not merely the oily part of the chyle or of the food; but is a new substance, or a new combination of the principles or elements, which is made probably in the secretory organs of the adipose membrane: the form of oil being made use of by Nature in preference to any other for the nutritious substance of the body, from its being the least liable to putrefaction, and from its containing the greatest quantity of nourishment in the least bulk. This circumstance was clearly proved by my valuable and ingenious friend the late Dr. Stark, who, in a course of curious experiments, made by weighing himself after living for some time on different sorts of food, discovered that a less quantity of suet was sufficient to make up for the waste of his body, than of any other sort of ordinary food; and that, when compared with the lean part of meat, its nutritive power was, at least, as three to one.

I MAY here add another circumstance that occurred to me when I first thought on this subject, which is, that since we believe the oil, or animal fat, is re-absorbed from the adipose membrane to serve for nourishment to the body; and as some of the patients (whose cases have been related above) could not take food, the re-absorption therefore of this oil might not be so much the cause, as the effects of the disorder under which they then laboured: or, in other words, that upon some defect in the digestive organs, the powers of nature drew from their magazines of oil in the adipose membrane, a supply of that fluid then perhaps necessary for the use of the body. In order to clear up this point, I thought it would be a satisfactory experiment, to compare the *serum* of the blood of animals at different periods after feeding them. For, if the re-absorption of the oil was merely to make up for the want of other food, or, if the *serum* was white merely

merely from a greater quantity of oil being taken up in order to supply the wants of the body, then the *serum* ought to be whitest in the animal kept longest without food, or whose body was most in want. And as I had found that geese had very commonly this white *serum*, though their chyle was transparent, I chose to make the experiment on them. I therefore took two of them that were very hungry, and feeding both of them with oats, one I killed four hours after, when I knew a part of the oats were undigested; and upon examining the blood, I found the *serum* whitish, and full of small globules; on its being suffered to stand a little time, the white part ascended to the surface like a cream. The other was killed forty-eight hours after eating, when its stomach was found empty, and the *serum* of its blood quite transparent, and without any cream rising to the surface, or any appearance of small globules, when examined with the microscope. Now, this experiment seemed

seemed to me decisive, and to point out clearly, that the whiteness of the *serum* was not occasioned merely by the body being in want of food, and therefore, drawing the oil from its magazines; because here the animal most in want of food had its *serum* least white; but was occasioned by the fat's being re-absorbed faster than it was used (from its place being supplied by the fresh chyle) and thence was accumulated in the blood-vessels, so as to give whiteness to the *serum*. And from the same observation it likewise appears probable, that the great re-absorption, and the accumulation of the fat in the vessels of the plethoric patients above-mentioned, was the cause of their want of appetite, and of their other complaints, and not the effect of them.

May not therefore a too great re-absorption of the fat, and its accumulation in the blood-vessels, be now admitted as the cause of one species of a *plethora*?

AND may it not likewife be ufeful in fome complaints of the ftomach, to attend to the whitenefs of the *ferum?* For, although fat be a fubftance little liable to difeafe, yet it may perhaps be fometimes fo vitiated or may fo incommode nature, that fhe may be obliged to take it up from her magazines, and to ufe it, or to throw it out of the body. Whilft this is doing, a ficknefs of the ftomach, and want of appetite, may be indications of fulnefs; and therefore, inftead of wanting remedies to ftrengthen the ftomach, may require bleeding, and other evacuations.

APPENDIX,

APPENDIX,

RELATING TO

THE DISCOVERY

OF THE

LYMPHATIC SYSTEM

IN BIRDS, FISH, AND THE ANIMALS
CALLED AMPHIBIOUS.

BEING

A Vindication of the AUTHOR's Right to thefe Difcoveries, in Oppofition to the Claim of Dr. ALEXANDER MONRO, Profeffor of Anatomy in the Univerfity of Edinburgh.

APPENDIX.

AN account of the Lymphatic System in Birds, Fish, and Turtle, was given to the Public in the Philosophical Transactions, vol. lviii. and lix. for which communications the Royal Society has since honoured me with their gold medal. These discoveries Professor Monro claimed, in a letter read before that most respectable body on the 19th of January 1769; and has since persisted in that claim, in a pamphlet called, A State of Facts, &c. printed at Edinburgh 1770. Now, as both that letter and the pamphlet must of course have been seen by many who know not what can be urged against them, I think it but a duty I owe my own character, to lay before

the

the Public the proofs I have collected of their infufficiency to procure Profeffor Monro the credit of having anticipated me in thofe difcoveries; and, I hope, that although in doing this I fhall trefpafs on the time and patience of the reader, yet it will be fome excufe for me, that I had endeavoured, as much as could be expected on my part, to fettle the difpute without troubling the Public with it.

As Profeffor Monro has, in this pamphlet, not only endeavoured to vindicate his claim to thefe difcoveries, but has likewife cenfured me on account of a paper on the *emphyfema*, it is neceffary, before I come to the controverfy about the Lymphatics, that I fhould relate what has paffed between us on that occafion.

In the third volume of the Medical Obfervations and Inquiries, is publifhed a paper on the *emphyfema*, in which I propofed the operation of the *paracentefis* of the thorax, to let air out of the cavity of the cheft; which air I endeavoured to fhew was the caufe of the worft symptoms attending that difeafe. Not long after this, I was informed that Dr. Monro had declared publicly, he had mentioned that obfervation in his lectures, both at the time, and before I attended them, which was in the winter 1761, and complained, that I had omitted doing him juftice in this particular.

WHEN I heard this, I made inquiries of fome of his pupils, who I found had taken notes at his lectures, and by two of thefe gentlemen I was favoured with excerpts from their notes, which convinced me that he had anticipated me in propofing that improvement. I then determined to let him know, that my omitting to mention his name on that occafion was entirely owing to my ignorance of his claim. This I was the more defirous of doing, from having heard that

that he had exclaimed againſt me with ſome acrimony, on the ſuppoſition that I had got the hint from him, and was conſcious of it; which being far from the truth, I determined to ſhew him in what manner I had really made the obſervation, and thereby ſtop his exclamations. I determined likewiſe to ſhew him that I was deſirous of giving him the credit of having had the idea before me, and thereby to prevent all diſpute about the matter. The following is a copy of the letter which I ſent him on that occaſion.

SIR,

BEING informed that you have publicly complained of me, " for having, " in a paper printed in the third volume " of the *Medical Obſervations* and *Inquiries*, omitted doing you the juſtice " of mentioning your having propoſed " the operation there recommended, in " the ſame circumſtances, long before;" and as I am confident I deſerved not to be complained of on that account, I have
taken

taken this opportunity of stating the manner of my making the obfervation, and at the fame time of letting you know, that fince I have learnt that you likewife had made it, I am willing to do you juftice. The thought firft occurred to me in reading Mr. Chefton's Pathological Inquiries and Obfervations, in which he gives a cafe of the *emphyfema:* this cafe is told in fuch a manner, that I think it is hardly poffible any unprejudiced perfon fhould read it and not be convinced, as I was, that the caufe of the principal fymptoms was air in the cavity of the cheft. Mr. Chefton himfelf, in relating that cafe, came as near making the obfervation as poffible †. From this hint I profecuted the fubject, as is

---

† I have fince been informed by Mr. Chefton, that I had mifunderftood his meaning, when I concluded that he explained the worft fymptoms of the *emphyfema* on different principles from what I have done, and that he meant to attribute them to air in the cheft, and therefore, that the obfervation was fufficiently made out in his paper; which fee in his Pathological Inquiries and Obfervations, p. 7, and 8.

mentioned in that paper; and before I published it, I consulted every author I could easily procure, who I thought was likely to treat of the subject. And I certainly should have done justice to any that I found had anticipated me, and should not have avoided the opportunity of doing you the same justice. But I knew not, at the time of that publication, that you had ever given the least hint on that subject. About the middle of last summer I was told by a gentleman from Edinburgh of your manner of treating me, at which I was not a little surprised, as I was not conscious of having given you the least cause of complaint. But having since learnt, from other gentlemen who attended your lectures before the time of my publishing that paper (and who, at my request, consulted their notes) that you had really mentioned it, I cannot now doubt that you had made the observation before me. At the same time I must assure you, that to suppose I knew it at the time of

publishing

publishing that paper was doing me injustice. Your accusation, I presume, is founded on the supposition of my having heard you deliver the observation at your lectures, when I had the pleasure of attending them. But I do assure you, that if I ever heard the least hint on that subject, either from you, or from any other person, I had not any remembrance of it at the time I wrote that paper. You are not, indeed, the only person who, as I now find, has anticipated me: the author of the Monthly Review for last June * says, he had long had the same idea, and that he mentioned it in his account of Mr. Chefton's book. But of this too I assure you I was ignorant, when I wrote my paper. What must give farther conviction to any unprejudiced person of my ignorance of your having made the observation, is this: I first mentioned it in a paper which I read to a private society, in which were

* See Monthly Review for June 1768, p. 446.

present many gentlemen that had attended your lectures, and yet all these gentlemen expressed themselves pleased with the observation, as new and interesting, and not one of them gave the least intimation of their having ever heard it before. And yet those gentlemen are as likely to remember any observation which tends to the improvement of physic or surgery, as any I know. I shall mention their names, to justify my good opinion of them; Drs. Stark, Parsons *, Saunders †, Pepys ‡, and Ruston §. The observation was likewise mentioned in another society of young gentlemen, and also in a public hospital, where many, who had been your pupils, heard it, and yet no person

---

* Professor of anatomy at Christ's-Church, Oxford.

† Physician to Guy's hospital.

‡ Physician to the Middlesex hospital.

§ Some of these gentlemen attended Dr. Monro's lectures about the same time with myself, the others since.

ever

ever told me before I published that paper (which was almost a twelvemonth after I had first mentioned the subject), that you or any other person had ever anticipated me. However, this I relate only to shew I was ignorant at that time of your having made the observation. But now I know that you had, I have not the least unwillingness to acknowledge it, and to do you justice in any future publication. At the same time that I justify my own conduct, give me leave to say, that your manner of treating me (if fairly represented) was not so civil as might have been hoped for. When you complained of me, 'tis a pity you had not likewise hinted there was a possibility of my being ignorant of your having had the idea. You might perhaps too without impropriety have hinted, that should it come to my ears that you had anticipated me, I might possibly be capable of such an exertion of candour as to acknowledge it. But, to have done with suppositions, this at least I am sure of, that though I may be as covetous of fame as moſt people, yet I am incapable of taking any unjuſtifiable methods of acquiring it.

I am, Sir, &c.

Dec. 31, 1768.

THIS letter, Profeſſor Monro could not but acknowledge, "ſufficiently ſatiſ-"fied him in having ſecured his title, as "the firſt who had propoſed that im-"provement." Yet ſo unfair an account does he give of it in his State of Facts, that he only ſays, I acknowledged, that "I could not doubt he had made the ob-"ſervation before me; but the farther "particulars of it (he adds) it is needleſs "to trouble the reader with, ſince as "much as is neceſſary of theſe will be "ſufficiently underſtood, from his letter "in anſwer" to me, which, ſurely! is not the caſe; for it no-where appears in his letter, that, beſides mentioning my conviction of his having anticipated me, me, I had likewise promised to do him justice in a future publication. Nor does it appear in his letter, that I had in mine shewn how little probability there was of my having got the idea from him. These the reader may perhaps think Professor Monro ought to have declared, in justice to me. For what more could be expected of me, seeing I had by accident hit upon an observation, which, as it happened, he had made before, than to acknowledge the priority of his title, as soon as I knew it, and to put that letter into his hands by which he might always be sure of securing to himself what was his due. But Professor Monro says, it was unnecessary to give a fuller account of my letter. But why was it so? Not surely in justice to me, nor for the satisfaction of the reader.——Nay, so far is Dr. Monro from doing me justice on this occasion, that he even intimates I rejected tapping the chest with a *trocar*, because it happened to be his method, as

if

if the same was not the method recommended by many of the writers on the subject of the *paracentesis* of the thorax, for the cases in which they advise that operation, to whose method I alluded, and not to his, which I then knew nothing about.

Next, as to the discovery of the Lacteals and Lymphatics in Birds, Fish, and the animals called Amphibious; of these an account was laid before the Royal Society on December the 8th, 1768. I was present when it was read, and had afterwards some conversation on the subject with Dr. Donald Monro, who, as appears by the sequel, informed his brother, the Professor, of what I had done. Not long after this, I again saw Dr. Donald at St. George's Hospital, and he then told me, that the *Lymphatics* and *Lacteals* in those animals * had

* Meaning Birds and Fish.

been

APPENDIX. 171

been difcovered by his brother eight years ago, as he now learnt by a letter from Edinburgh, a part of which letter he was to fhew to every body, and which was already given to be read before the Royal Society. When I was informed of this, I was aftonifhed, as I remembered to have heard the Profeffor, fince that time, declare that they were not difcovered. Befides, I had a note taken from his lectures within two years of his making this claim, in which was a fimilar acknowledgment *. I was convinced therefore that he had no title to thefe difcoveries. Upon which I laid before the Royal Society my reafons for that conviction, in a letter to one of their fecretaries. Of this letter I fhall give the reader an account, but fhall firft lay before him a literal copy of Dr. Monro's claim.

* That note is now printed below, p. 176.

Copy

Copy of Dr. Alexander Monro's *claim, &c. read before the Royal Society,* Jan. 19, 1769.

" Above four years ago (fays he) I injected the lacteal veffels of a turtle, or fea-tortoife, with quickfilver, after injecting the artery and vein with wax, and have fhewn this inftance of the veffels in the oviparous animals every year in my college, and had a drawing made of it two years ago by Dr. Palmer, a copy of which I have fent inclofed, engraved by Donaldfon.

" I likewise, eight years ago, mentioned thofe veffels in Fowls and Fifhes, which I had feen, but not injected."

Here then is an affertion about the veffels which I had difcovered, that is far from being equivocal. For here he affirms,

affirms, that he really had seen them eight years ago, nay, that he had even mentioned them to others. This letter too was sent immediately after he heard that I had laid before the Society an account of those vessels in Birds and Fish*. It could not therefore be meant merely to inform the Society that, now seeing Mr. H. had discovered the Lacteals and Lymphatics in Birds and Fish, he likewise had the pleasure of shewing them that he (Dr. Monro) had discovered them in the turtle. This, I say, could not be his meaning; for if it was, why did he send his letter so precipitately? Why did he not send a description of those vessels in the turtle, in order to make his letter worthy their notice? And why did he say he had seen them in Birds and Fish? The Society, he knew, wanted not his testimony to prove that Birds and Fish had them. What then could he mean by it, but to claim the discovery?

* As he acknowledges, State of Facts, p. 4.

As there could be no doubt that Professor Monro meant this letter as a claim to these discoveries; neither had I any doubt but I should, for the reasons above-mentioned, prove that he had no right to them. In order therefore to prevent the prejudices that might arise against my papers, from his being believed to have anticipated me in these discoveries, I wrote a letter to one of the secretaries of the Royal Society, in which I first shewed, that I had seen the Lacteals of the turtle about a year before him, and then, when I came to speak of those vessels in Birds and Fish, against the probability of Dr. Monro's having anticipated me in these discoveries, I made use of the following arguments.

1st, His not having, by his own confession †, injected them, which he certainly would have done, in order to complete the discovery. To which I

† In his claim; see above, p. 172.

observed he had the strongest motive, both from his knowing the importance of the subject, and from his having unfortunately declared, in the 57th page of his Anatomical and Physiological Observations, printed at Edinburgh, 1758, " that, " after a considerable number of expe-" riments which he had made, he was " convinced, that neither Birds, Fishes, " nor oviparous animals in general had " either Lacteals or Lymphatic Vessels." After which declaration, I conceived it improbable he should patiently wait eight years without injecting them, especially as I had found it an easy matter to inject them, when once they were discovered. And I added, the probability was, that if he had *seen* those vessels, he would have hastened to inject them, and to complete the discovery, were it only to prevent another person's doing it, and thereby acquiring the reputation of having done what he himself had *in vain* attempted by such a considerable number

of

of experiments as *were sufficient to convince him*, that such vessels existed not in those animals.

2*dly*, I SAID, his claim to the discovery of those vessels, by affirming he had seen them eight years ago, was contradicted by public declarations made after that time; for he had, since, acknowledged in his lectures, *that he had sought for them in vain by a variety of experiments.* And even so late as within these two years †, he declared likewise in his public lectures, " that the Lymphatic Sys-
" tem was supposed to take place only
" in men and viviparous animals, and
" by analogy in those fishes placed by
" Linnæus amongst the *mammalia*, and
" how far was their just extent (he said)
" was not certain, but that he had found
" them in some amphibious animals, as
" in the turtle." These declarations, I

† My letter being dated Jan. 10, 1769.

observed,

obferved, were inconfiftent with his claim to the difcovery.

BESIDES ufing thefe arguments, I promifed the Society I would hereafter produce unqueftionable proofs of the invalidity of his claim, having by this time found, that the Doctor, fortunately for me, had exprefsly acknowledged in his lectures, *that he had fought for them in vain,* almoft every year fince the time that he now pretends to have feen them.

DR. MONRO being informed of thefe proceedings, fent me his letter, dated June the 8th, which he has fince printed in his *State of Facts*. But that letter appeared fo confufed, that I knew not what to make of it. Sometimes I thought it was meant to prove that he had difcovered thofe veffels, agreeably to his affertion read before the Royal Society: but this I foon after fufpected could not be be the cafe, becaufe, after relating all his facts and experiments, he concludes, *not* that he had difcovered them, *but only* " that he had feen what he ftrongly " *fufpected* to be lacteals in thofe ani- " mals" *(viz.* Birds *)—And, " that " (from preceding experiments) he was " *perfuaded* that Birds were provided " with lacteal veffels, and confirmed in " this *opinion,* by having injected them " in one of the fame oviparous clafs, " the turtle †."——Or, in other words, Dr. Monro claimed a difcovery, by telling the Royal Society that he had *feen* thofe veffels in Birds eight years ago, which he now proves, by fhewing he had only *fufpicions* about them, and an *opinion* that Birds had them, becaufe Turtles had them.

AT other times, I thought that pof- fibly, after finding what his pupils tefti-

---

* State of Facts, p. 21.
† Ibid. p. 23.

fied,

fied, he might now be convinced he had imprudently claimed those discoveries, and might intend this as a sort of an acknowledgment (tho' an aukward one) of my being the first who had seen those vessels. But the vindictive stile of his letter convinced me it was not meant as an apology †, as likewise did the stile of a short note written on the cover of that letter, of which note the following is a copy.

To Mr. Hewson, &c.

SIR,

When you have read the inclosed, you are very welcome to write such remarks on it, as to you, or to your friend Dr. H———r, or to any of his friends,

---

† As for instance, where he talks " of (his) " making all the allowance I require for my natural, " or (says he) I should rather call it, unnatural " imbecility of memory." This passage is altered in his *State of Facts*, p. 22.

such as Dr. ——, &c. may seem proper; only when you have done, I think you ought to shew it to all those societies, physicians, and students to whom you have made free with my name; or, if this task should not suit your disposition, or be irksome to you, after the great fatigue you have taken about me already, please to let me know this, and I shall take that trouble on myself.

I am, Sir, &c.

(Signed) ALEX. MONRO.

Edinburgh, June 24.
1769.

This seemed clearly not to be the stile of one who was sensible of his error, and was apologizing for it; and convinced me, that Professor Monro intended that letter as a proof of his right to those discoveries. However, not to be positive that

that I had hit upon his meaning, I determined, before I laid any thing before the public, to afk an explanation of that letter. For this purpofe I wrote to him on the 15th of July, and defired him to tell me " whether he meant it as a " proof of his right to thofe difco- " veries"——or " whether he meant by " it to give up to me the right to them." And as I had found him in that letter wandering from the fubject, and inftead of concluding that he *really* had feen thofe veffels, concluding *only*, that he had feen what he *fufpected* to be the Lacteals in Birds——And again, that he was *perfuaded* Birds were provided with thofe veffels—but no where faying, that he had feen what he *knew* or *could prove* to be their Lacteals, which alone could give him a right to the difcovery: I therefore told him, " that to avoid for " the future all wandering from the " fubject, I fhould ftate the difpute as " it appeared to me;" and then I faid, that " it was he who began it, for, on

" hearing

"hearing that I had difcovered the Lymphatic Syftem in the three claffes of oviparous animals, he had fent a letter, which was read before the Royal Society, and was to be fhewn to every body. In this letter he afferted, that he had difcovered the Lacteals in a Turtle about four years ago, and in Birds and Fifh eight years ago; and that he even mentioned thefe difcoveries to others." Thefe affertions (I added) were conftrued a claim to the difcoveries I had made. With this letter I likewife fent him a copy of mine, which had been read before the Royal Society. By thefe means I thought I fhould either keep him to the fubject in queftion, or, if he fhould again wander, the reader would be convinced it was not my fault, but his own, that he *now* knew not what he had *then* afferted.

This letter, however, had no effect. I therefore wrote to him again, hoping he he might now be convinced that his claim was ill-founded, and might therefore be induced to retract it, instead of obliging me to prove to the world its invalidity. The following is a copy of the letter which I sent him on that occasion:

SIR,

It is now above six weeks since I wrote to you, desiring an explanation of your letter of the 8th of June. As you have not given me that explanation, I have now taken up the pen to inform you, that agreeably to your own desire, and in order to justify my conduct towards you, I am commenting upon that letter which you sent me. My comment would be more to the purpose, were I always sure I understood you, but if that satisfaction should still be denied me, I must proceed as well as I can, and I must say, that if I should mistake your meaning, it will not be wilfully, since you might, by an anſwer, have cleared up all ambiguity. I cannot help regretting, that this diſpute ſhould ſubſiſt between us, both on my own account, as I think it hard to have the trouble of proving my right to diſcoveries which are certainly my own, particularly as it takes up that time which I hoped to employ to a better purpoſe; and I likewiſe regret it on yours, ſince, in order to maintain my right, I ſhall be under the neceſſity of producing ſome facts and teſtimonies, which, in my opinion, cannot but lead to concluſions very unfavorable to your reputation. And as I ſhould be ſorry that one of my firſt attempts to lay the foundation of my own character, ſhould be attended with circumſtances which may hurt yours, and really wiſh to avoid it; I therefore ſtill hope that this diſpute may be ſettled in a more eaſy manner. You muſt, I think, be now convinced, that in claiming theſe diſcoveries you have injured me, and cannot be at a loſs to know what might be be expected from you on such an occasion. But if instead of doing me that justice which might be expected from a man of candour, you treat this letter likewise with silence, then justice to myself requires that I should no longer delay producing such proofs as I possess of your having no right to these discoveries, and shewing them to the very respectable Society, to which I have promised them; or to such *physicians, students, &c.* as may have heard of your claim; without regretting much that those measures which I take to maintain my right, may perhaps affect sensibly the character of a man, who having first injured me and afterwards had his error pointed out to him, was incapable of candidly acknowledging it.

I am, Sir, &c.

(Dated)
Sept. 9, 1769.

In answer to this letter, he sent me one dated Sept. 30, in which, instead of answering my questions, he evades them, concludes as vaguely as in his former; and when he speaks of his assertion read before the Royal Society, alters its sense, qualifying the alteration with "*to the best of his recollection*" denies he was mistaken in claiming these discoveries. And, what is still more remarkable, accuses me of having made conclusions injurious to him, "*by arguments, weak, inconclusive, not real, but feigned.*"

It was evident, from such a letter, that he would not embrace the opportunity I offered him, and avoid a dispute, by acknowledging his mistake, and retracting his claim. I therefore no longer hesitated to print the proofs I had collected of his not having anticipated me; and though I had once intended to make some remarks on his letter of the 8th of June, as is mentioned above, yet I afterwards determined

determined to omit these, since the testimonies of his pupils alone sufficiently proved that he had no right to those discoveries. By these means I reduced the publication to half a sheet of paper: in which I first gave an account of his claim made, by saying " *he had seen those* " *vessels* EIGHT *years ago*;" then I mentioned, as arguments against its *validity*, that I had myself heard him since that time declare, " *he could not find those* " *vessels*," and that, besides, I had a note taken at his lectures by a gentleman within two years of his making it, which contained a similar declaration *, and afterwards

---

* Dr. Monro has misinterpreted this passage. He supposes I meant that this note was taken two years before 1762, whereas I meant two years before 1768, the time when he claimed these discoveries, and when I wrote my letter to the secretary of the Royal Society. The note itself is printed above, page 176.

Dr. Monro has taken notice of another inaccuracy, that is, where I had said, " He asserted that " he

afterwards I said that I had written to such gentlemen as I knew had, as well as myself, attended his lectures within these eight years, desiring them to consult their notes, and to let me know what Dr. Monro had said as to the existence, or non-existence, of the lacteals and lymphatics in these animals; without mentioning the dispute between us, or any opinion I had formed, that they might be unprejudiced to either party. And that such of these gentlemen as had taken notes sent me excerpts from them, which, as I had suspected, agreed with what I had myself heard the Professor say upon that subject.

THAT Dr. Haygarth, physician at Chester, had sent me the following pas-

---

" he had anticipated me in these discoveries;" instead of saying " that he claimed them by asserting " he had injected, &c." The former Dr. Monro seems to allow to be what he meant, but not *exactly* what he said. It was therefore a small inaccuracy. But his claim is now printed *verbatim*. See above, p. 172.

fage from the notes which he had taken in 1764.

"Fowls, and some Fish (says Dr. Monro) have not lacteal vessels that that we can see; they have no conglobate glands in the mesentery; perhaps they (the lacteals) don't run into each other, but into red veins, and hence never are so large as to become visible."

"This note," adds Dr. Haygarth, "was taken in 1764, and if Dr. Monro had changed his sentiments on this subject in the year 1765, I should certainly have taken notice of so remarkable a circumstance."

That, from Mr. Orred, surgeon at Chester, who attended Dr. Monro's lectures during the winter 1765-6, I had learnt, that Dr. Monro, when he spoke of the anatomy of a cock, declared: "he
" never

" never saw, or observed any *glandulæ*
" *vagæ*, or lacteals, but had seen lym-
" phatics in the neck, ending in the
" jugular *."

That, from some notes, said to be copied from those taken by Dr. Taylor of Reading, in the winter 1765-6, I had procured the following excerpt; and Dr. Taylor, on being requested to consult his own copy, had acknowledged it was a just one.

" The lymphatic system (says Dr.
" Monro) is said to take place only in
" men, and viviparous animals, and
" from analogy in those fishes placed by
" Linnæus, under the class of *Mam-*

---

* It may be necessary to mention here, that the dispute between Dr. Monro and me is, *who first discovered the lacteals of birds?* for as to the *lymphatics* in their necks, (mentioned in this gentleman's note,) these we both allow, were discovered by Mr. John Hunter about ten years ago.

" *malia:*

" *malia* *: how far is their just extent
" is not certain, but we have found
" them in some amphibious animals, as
" in the turtle †.

" It is said that this system is want-
" ing in oviparous animals; but this is
" not universally true; for we men-
" tioned, that we found them in a
" turtle, and they would probably be
" found in other orders and *genera*, if
" properly examined. But admitting
" that they are not demonstrable there,

---

* In the excerpt it is *amphibia*; but it is evident from the sense, and from comparing it with the other notes, that it should be *mammalia*.

† As to the *lacteals* of the turtle, there is no doubt but that Professor Monro and myself have both discovered them. He, in the summer 1765; I before that time, viz. in the autumn 1763; when I took a short description of those vessels, which is published with my paper on the lymphatic system in birds, in the 58th volume of the Philosophical Transactions.

"it doth not follow that they are wanting; for, perhaps, they may run
only a little way, and terminate in red
veins."

That Dr. Maddocks, phyſician to the London Hoſpital, had favoured me with the following excerpt from notes which he took at Dr. Monro's lectures, in the winter 1765-6.

"Lymphatics are found in viviparous animals, and therefore, I pre-
ſume, in the whale, which is of this
kind. They are not to be found in
oviparous ones, fiſhes, nor the *amphibia:* this is the common doctrine.
*I will not ſay how far they may be
found in ſome birds,* but I have found
them in ſome of the amphibious animals, as in the turtle, running along
the root of the meſentery."

That

APPENDIX.

That Mr. Hull, furgeon at Stevenage, had fent me the following excerpt from his notes, taken about four years ago.

"I never could, to this day (fays Dr. Monro) find a fingle branch of a lacteal in the abdomen of fowls, nor any lacteals, or glands of the conglobate kind in the mefentery, notwithftanding I have made experiments with that view very often. I kept fowls twenty-four hours without food, then fed them with bread foaked in milk, and tinged it by turns with blue, madder, and faffron, and afterwards opened them at feveral different times, in order to difcover the lacteals, but all without fuccefs. Yet, perhaps, the lacteals may be in fowls, though not demonftrable." This, adds Mr. Hull, I will anfwer for being *verbatim*, or nearly fo, as Dr. Monro delivered it in the anatomical hall, at

Edinburgh, on the 13th of February, 1765 *.

THESE paſſages, I added, were ſufficient to ſhew how little right Dr. Monro had to theſe diſcoveries. Beſides, I ſaid it was a ſtrong argument againſt him, that in the letter I had received (which he has printed in his State of Facts, p. 8.) he could not, after relating all his experiments and obſervations, conclude he had *really* ſeen thoſe veſſels as he had told the Royal Society; but in one place he ſays, " that, from the preceding " experiments, &c. it is evident that he " had ſeen what he *ſuſpected* to be the " lacteals in birds." And in another, " that he was himſelf *perſuaded* that

* If the reader will take the trouble of comparing this note with Dr. Monro's own account of his experiments, (State of Facts, p. 12.) he will be convinced how accurately this gentleman muſt have taken notes.

" birds.

" birds were provided with lacteal vef-
" fels, and confirmed in this *opinion*, by
" having now injected them in one of the
" fame oviparous clafs, the turtle †."

Such conclufions appear merely eva-
five, and never can be confidered as
proofs of his having difcovered thofe
veffels before I did, agreeably to his affertion read before the Royal Society, and
fince repeated in his State of Facts.

The half-fheet of paper, containing
thefe arguments and teftimonies againft
Dr. Monro, was printed Dec. 1, 1769,
and was given to fuch gentlemen of my
acquaintance as had heard of his claim,
and a copy of it was fent to the Doctor.
Upon receiving which, he publifhed his
State of Facts; but, what is fingular, he

---

† This argument was repeated in a note, to prevent Dr. Monro's writing; as if the difpute between us was, *who firft had fufpicions about thofe
veffels*, inftead of *who firft difcovered them.*

has attempted his juſtification without taking proper notice of theſe teſtimonies againſt him; as if he could be juſtified whilſt they remain unanſwered. And in this State of Facts, in ſpite of thoſe teſtimonies, he repeats to us, " that long " before 1762, he obſerved blueiſh veſ- " ſels in the meſentery (of birds) which " he *judged* to be lacteals, and *had men-* " *tioned as ſuch* in his lectures *." And again, " that about the years 1759-60, " he had ſeen collapſed blueiſh veſſels, " which he *concluded* lacteals, &c." †. What ſhall we ſay of this?

NAY, Dr. Monro has, upon this occaſion, even ventured another aſſertion, viz. " that the notes of his pupils taken " for three years before 1762, will be " found to prove, that he then taught " the direct contrary" of what I have brought theſe teſtimonies to prove he has

* State of Facts, p. 4.   † Ibid. p. 27.

ſince

since taught ‡. Now, surely this is very improbable; and Dr. Monro should have adduced some testimonies to prove it. But supposing it were true, it would lead to a conclusion unfavourable to him. It would shew, that he must have misled either the one set of gentlemen or the other;—for he says, he told the first he had seen the lacteals—the last prove he has since taught them that he *never could see* those vessels.

THE reader, I fancy, by this time thinks with me, that Professor Monro's claim deserves no more of our attention. But, as he has printed some excerpts from his OWN book of notes, with the parade of having them authenticated, as if they contained the discovery, notwithstanding the above-mentioned proofs of his having acknowledged repeatedly since he wrote them, that he never could find

‡ State of Facts, p. 26, in the note.

those vessels, I shall next, therefore, make some remarks upon his notes.

To begin with those relating to the turtle. He discovered its lacteals in the summer 1765. I had seen them before that time, viz. in the autumn 1763. Besides, I have since injected and traced out the whole system*; he does not even pretend to have done so: it is, therefore, not difficult to determine, who was the first discoverer, and who has carried the discovery farthest.

Next, as to the lacteals in Fish. To prove that he had found those vessels eight years ago, he tells us, that in a note taken from the dissection of a skate on April 24, 1760, he has said, " He " had discovered a whole system of " lacteals and lymphatic vessels running " towards the heart, on the left of, and

---

* See Philosoph. Transact. vol. lix.

" above

APPENDIX.

"above the *vena portarum*, and from these the auricle of the heart was blown up. They are proportionally larger, but have fewer valves than in man †." Now, I will take upon me to say, there is nothing in this note which proves whether he had inflated a lacteal or a vein. For what he says of the situation of the vessels, and of his blowing up the heart, is equivocal: the only part of the note which appears to characterise the lacteals, is in reality a mistake; that is, where he says *they have valves*. But the lacteals on the mesentery of a skate have no valves, and injections pass readily from the large to the smaller branches. And what is even more to the purpose, although it appears, from his calling what he saw *lacteals* and *lymphatics*, that he had at that time some suspicions about them. Yet I am persuaded he has since changed his opinion; and this I think is evident, even from the

† See State of Facts, p. 12.

manner in which he speaks of his experiments made the year after. For, says he, " I have dissected this year (1761, in summer) eight skates, and about a like number of cods and codlings, but without being able to observe by dissection, or to inflate any like to lacteal or lymphatic *glands*—I find indeed (he adds), that blowing backwards in the meseraic veins, the intestines and the cellular substance between their coats are inflated; but this is no direct proof of branches of red veins absorbing, as these veins may be burst, or the air may have first entered the arteries *." Now, this surely is not the language of a man who had seen the lacteals, but of one that was seeking for them. Had he found them, he certainly would have mentioned it in this note, but he avoids the subject entirely, and only says he could not find the *glands*, thereby leaving us to suppose that these

* State of Facts, p. 12.

APPENDIX.

diffections were made for the *glands only*, after having difcovered the *veffels:* which is highly improbable, fince, by his own confeffion †, he did not inject the veffels, which he knows well enough is the beft way of determining whether the glands exift or not; and *one* experiment in this way would have been more fatisfactory than his *eight*, or than eight hundred made by diffection only ‡. Add to

---

† See his claim above, p. 17?.

‡ If the reader fhould happen not to be well-acquainted with this part of anatomy, he may not fee all the force of this argument, which will be fatisfactory to anatomifts; for it is a fact admitted amongft them, that the mefenteric glands are placed only in the courfe of the lacteals, fo that the lacteals muft pafs through thefe glands in their way to the heart. The readieft method therefore of difcovering the glands, after having feen the veffels, is by injecting thofe veffels; for the injection, in its way to the heart, muft diftend the glands, and make even the fmalleft of them vifible. The veffels feen by the Doctor feem to have been very large; can it be fuppofed then, if he had been convinced they were lacteals, that he would not have injected them,

to this, he would not, I think, if he had now seen the lacteals, have taken up his time with trying whether the red veins did the office of absorption for them, as he seems to have done by blowing into these veins. Nay, I will go farther, and will take upon me to say, that it is probable he was in these last dissections convinced he had been mistaken in what he took the year before for lacteals and lymphatics. This I think evident, both from the notes abovementioned, and from his manner of treating the subject since that time. For, if he thought he had seen those vessels, he would doubtless have used this discovery as an argument against absorption by the common veins, as he has since used that in the turtle. But it appears, from the notes of his pupils †, and even

---

them, and thus have determined whether there were glands or not, by one experiment, instead of tediously dissecting sixteen Fish?

† See his pupils notes above, p. 189, &c.

from his own account of those arguments †, that he has not done so. And again, had he thought he had discovered those vessels, he would not have acknowledged in his lectures since that time, that they were not yet known to exist ‡. He has therefore aggravated the impropriety of his conduct in claiming these discoveries, by the disingenuity of sending such notes as proofs of his claim.

And lastly, as to the lacteals in *birds*, he tells us, " that, in 1758, he re-
" marked a vessel making an arch on
" the mesentery of a cock, which at
" first he believed to be a trunk receiving
" the lacteals, but not being able to in-
" ject it on trial, he conjectured to be
" rather a nerve." And afterwards, in April 1759, " he observed in a cock

† See State of Facts, p. 16.

‡ See the notes of his pupils above, p. 189.

" what

" what *looked like* lacteal veſſels col-
" lapſed, of a blueiſh colour, which
" ſeemed to terminate at the back-bone,
" &c.—Theſe he ſhewed to the ſtu-
" dents." And again, after relating
the manner of making his experiments
upon no leſs than twelve cocks *, he
tells us, " that in 1761 he obſerved, in
" the interſtices of the great arches of the
" red meſenteric veſſels, a pellucid net-
" work, ſome part of which ſeems to be
" compoſed of branches ſent from a
" large nerve, running parallel with the
" inteſtines, and nearer to them than
" where the trunk of the meſenteric
" artery ſends off its large branches;
" but although (ſays he) *I ſuſpect*
" *ſtrongly* there are here too numerous
" lacteals, and I even obſerve very ſmall
" knots, which I take to be analogous

* The experiments were made by feeding theſe birds with oatmeal and madder, oatmeal and rhubarb, &c. See State of Facts, p. 12.

" to

" to our mesenteric glands, yet I have
" not observed the above-mentioned
" kinds of food to make any odds in their
" appearance *, &c." And again, after
a variety of other experiments, he says,
" he could not observe more than above-
" described." Now, what is there in
these notes that can entitle him to the
discovery of the lacteals in birds?—Can
his seeing a blue plexus on the mesentery,
which at first indeed he suspected to be
lymphatic, but afterwards to be nervous,
and a part of which, he acknowledges,
was in his last experiments found to be
made of nerves, entitle him to it? or
can his discovery of these small knots,
which he takes to be analogous to our
glands, entitle him to it? Certainly not.
Birds have no lymphatic glands on their
mesenteries, as I have shewn †. Is it
not therefore plain, from these notes

* State of Facts, p. 12.
† Philosoph. Transact. vol. lviii. art. 34.

themselves,

themselves, that he had not discovered the lacteals in birds? Has he not repeatedly since that time acknowledged this in his lectures *? What shall we say then to his asserting that he had seen them eight years ago, and his laying before a respectable Society, and desiring his brother to propagate such an assertion? Or what shall we say to his persisting in it, or above all to his telling us, in his last publication (p. 26.) " that " in these notes the reader will find the " appearance of these vessels after death " really described † ?"

AFTER these notes follow some others, to prove that he had *argued* in favour of the probability of the existence of those

---

* See the notes of his students above, p. 189.

† Let me beg of the reader again to examine these notes, and then judge of the propriety of Dr. Monro's affirming they really contain a description of those vessels, when he has himself put it in the power of the reader to observe the contrary.

vessels

vessels in Birds and Fish; and a conclusion that he had *supposed* frogs might have them; and his *suspicions* that Birds might have them—And his *persuasions* that they must have them *(not because he had seen them, but)* because turtles had them. Which are nothing to the purpose, and ought never to have induced him to claim the discovery, or to say he (actually) had seen them. I cannot therefore think it worth while to take any farther notice of these conclusions.

It is indeed remarkable, that Dr. Monro could persuade himself he had any original merit *even* in entertaining such *suspicions* and *opinions*. More than one writer had suspected those animals had them, and that they themselves had seen something like them; for a proof of which the reader need look no farther than Dr. Haller's *Elem. Phys.* *. But, as those writers had given no proofs of

---

* Lib. ii. Sect. 3. & Lib. xxiv. Sect. 2.

their

their having discovered them, their *suspicions* and *opinions* past for nothing.

PROFESSOR Monro, not satisfied with claiming these discoveries, has even gone farther; he has intimated, in some parts of his book, that I might have learnt them, or a part of them, from him. As in page 4, where he speaks of my " giving an account of these vessels en-" tirely as my own discovery," this in page 6. he calls " broaching another " subject with him;" and complains of me " for passing in silence what I might " have heard him observe concerning it " when I attended his lectures."—How Professor Monro could pretend that I had learnt any thing from him on this subject, that ought not *for his sake* to be passed in silence, is astonishing. What could I learn from one who has repeatedly since that time acknowledged *he never saw* these vessels; *that they might be too short to become visible*; and who,

at

APPENDIX. 209

at the time I attended his lectures, faid, he could not find them, as I have already declared. But, as my teftimony will have more weight with the reader, when corroborated with that of a gentleman unconcerned in the difpute, I fhall next add a copy of fome notes taken by Dr. Morgan, now profeffor of medicine in Philadelphia, who attended Dr. Monro's lectures at the fame time with myfelf, and who, at my requeft, fent me the following excerpt, taken at his lecture upon the queftion, Whether the common veins abforb or not?

" Most authors (fays Dr. Monro)
" concurring in opinion, that fowls
" were deftitute of lymphatics, and not
" being able to difcover them myfelf, I
" was led to be of their opinion. I have
" already obferved, that where conglo-
" bate glands are found, there are
" lymphatics, and the converfe of this
" pro-

"proposition, namely, where there is no conglobate gland, there are no lymphatics. And there being no conglobate glands to be seen in the mesentery of fowls, nor in fishes, I judged these animals to be destitute of lymphatics; but Mr. John Hunter having discovered conglobate glands in the neck of a swan, put me on further search, and I then found them plainly in common fowls, *but never could find any lacteals* in their mesentery, though experiments were tried by means of coloured tinctures of various sorts, as of rhubarb, &c."

From this excerpt it is evident, that Professor Monro, when I attended his lectures, taught, as he has since done, that what he knew of the lymphatics he learned from Mr. Hunter, and as to the lacteals *he could not find them*; and this was in the spring 1762, the very year after the time when, according to his letter

letter read before the Royal Society, he should have seen these vessels, and mentioned them in his lectures: and finally, to complete the whole, he now complains of me for passing in silence what I might then have heard him observe concerning them.

Thus have I endeavoured to obviate the arguments in Dr. Monro's publication, and the reader must now, I think, see clearly, not only the impropriety of the Doctor's asserting his right to these discoveries, but the still greater impropriety of his persisting in that assertion.

Besides claiming these discoveries, Dr. Monro has, in his letters on the subject, treated me in a manner which I cannot pass unnoticed—Thus, he first gives the name of misinterpretation to my concluding from the notes of his pupils,

pupils, that he had not seen " what " he believed" the lacteals, and then adds:

" SHOULD we even suppose the above " misinterpretation venial, what must " the reader think, when he is told, " you was informed that a gentleman, " who had attended my lectures two " years at least before I injected the " lacteals of a turtle, that is, nearly " about the time you did, declared he " heard me then speak of having *seen* " *the lacteals in fowls*; and yet that " you continued to vent this injurious " supposition? That is, you must have " sunk this material information, since " it overturned the whole purport of " your story *."

Now, here is an accusation, which, were it true, would fall heavy upon me.

* See his State of Facts, p. 23.

But

But the case is this; I had indeed heard that a gentleman, who attended Professor Monro's lectures about the time I did, had declared he then understood the Professor had seen the *lymphatics* in Birds. And Dr. Donald Monro, when I saw him at St. George's Hospital, asked me, if I had not heard that this gentleman (mentioning his name, viz. Dr. James Blair, then in London, now in Virginia) had said, that the Professor had then mentioned his having seen the *lacteals* in Birds. I answered Dr. Donald, that I had heard something of the matter, but could conclude nothing from it (or to that effect). The reason was this; I knew from the testimony of my own memory, that the Professor had then acknowledged the contrary of his ever having seen the lacteals †. I knew the same from the testimony of gentlemen

† Dr. Morgan's note proves he did so, see above, page 209.

who had attended his lectures since. I therefore concluded that this gentleman had confounded his saying he had seen the *lymphatics*, with his saying he had seen the *lacteals*, which I thought might easily happen, as I never knew him take any notes. And upon receiving the Professor's letter, I wrote to Dr. Blair, and in his answer he acknowledged, " that although he had, indeed, for " several years, been under a general " persuasion that Dr. Monro had seen " the *lacteals* or *lymphatics* in fowls, " yet he had no note on the subject, and " a very confused remembrance of what " he had heard."

Similar to this accusation is the greater part of a letter which I received from Dr. Monro, in answer to two of mine. This letter is dated Sept. 30, 1769. The Doctor has not printed it, but I beg leave to take a little notice of it.

He

APPENDIX.

He begins it by altering the sense of his assertion read before the Royal Society, by the introducing the word *believed*, making it rather a doubtful than a positive assertion. He has done the the same in the beginning of his State of Facts, qualifying the alteration, by adding, *" to the best of his recollection, " and that he had not kept a copy of his " letter, not supposing it material to do " so *."* But surely this was not sufficiently qualifying it. If he did not know exactly what he had then asserted, why, before he defended it, did he not ask a copy from his brother, who, most probably, would keep it, *in order to shew it to every body*; or solicit that favour from the secretaries of the Royal Society where it was read, who, he might be sure, had preserved it, as they do every paper that is laid before the Society. And again, if he was not sure

---

* See State of Facts, p. 5.

of its contents, how could he now venture, in his State of Facts, *positively* to insist, in opposition to what I had declared, " that his first and last assertions " were exactly the same *." This at least was inexcusable.

NEXT, he repeats his vague inferences as in his letter of the 8th of June, " that " he had seen what he *suspected* to be " those vessels, &c." and afterwards, when he comes to speak of the conclusions concerning his claim, which I made in my letter read before the Royal Society, he says, " That he was almost " ashamed, on my account, to add a " plain corollary, that I must or might " have been conscious, that the injurious " conclusion with respect to him, which " I was labouring to impress on the " members of a respectable Society, was " drawn from arguments that were " weak, inconclusive, not real, but " feigned."

---

* State of Facts, p. 26, in the note.

" feigned *." Afterwards he tells me, " that he is glad, *on my account as well* " *as his own*, that I am at last really " ashamed of my letter." And he then finishes with the following passage: " Another unhappy mistake of yours " (says he) is, that you should not have " known, or rather perhaps misfortune " of yours, since you don't seem to have " known so much, that you should not " have been told, that your presuming " to draw the above conclusion concern- " ing any person who had the smallest " pretence to character, without pro- " ducing proof and absolute certainty of " its being true, was what you never " could be able to justify to any *gentle-* " *man*."

Now, when it is considered that Dr. Monro obliged me to act in the manner I have done, in order to secure my

---

* The conclusions alluded to in these passages are printed above, p. 175, and 176.

right, do not these passages appear very extraordinary?——But the reader, I believe, will excuse my not dwelling upon them. I shall therefore only add, that the *proofs* on which my *conclusions* were founded, being now laid before the public, to their judgment I willingly submit them, and that, with respect to Dr. Monro, I have nothing more to say, than that I hope, *for his sake as well as my own*, to see no more of his *claims*, his *assertions*, and his *conclusions*.

The END.

ERRATA.

P. 92, in the note, *for* Experiment 31, *read* Experiment 32. P. 98, in the note, for *in* read *of*.

PROPOSALS

FOR A COURSE OF

Anatomical and Chirurgical Lectures;

By WILLIAM HEWSON, F. R. S.

In this COURSE it is propofed,

Firft, To explain the Structure and Functions of the feveral Parts of the Human Body.

Secondly, To apply this Knowledge to the Cure of Difeafes, particularly fuch as require manual Operation; and to fhew the various Operations of Surgery, and the Manner of applying Bandages.

Thirdly, To examine the Structure of the impregnated Uterus, and its Contents; in order to facilitate the Study of Midwifery.

Fourthly,

Fourthly, To compare the Structure of the Human Body with thofe of Quadrupeds, Birds, Fifh, and Infects.

The Whole to be comprehended in about an Hundred Lectures, which may be attended on the following Terms:

> For the Firft Courfe, 3 Guineas.
> The Second, 3 Guineas.
> The Third, 2 Guineas.
> The Fourth, 2 Guineas.
> The Fifth, 1 Guinea.

After which each Gentleman may attend the Lectures as long as he pleafes without additional Expence: Or, for *Ten Guineas* paid at Once, each Gentleman may have the fame Privilege.

The Art of *Diffecting*, *Injecting*, and making Anatomical Preparations will be taught on the following Conditions:

For *Ten Guineas* paid at Once, each Student may have a Right to attend the Diffecting-Room as long as he pleafes; and to be taught the various Arts of Injecting, and making Anatomical Preparations, paying for the Subjects he ufes.

Or, for *Six Guineas*, any One may have that Privilege for Three Months.

But fuch as chufe only to attend a fhort Time, may diffect a fingle Subject for Two Guineas, and may have any Part of it injected at a moderate Expence.

Thofe who are defirous of merely vifiting the Diffections without operating themfelves, may have that Privilege for half the Sum that is paid by thofe who diffect, or may be perpetual Vifitors for Five Guineas; and for Three Guineas may attend the Diffecting-Room for Three Months.

For *Twenty Guineas* paid at Once, each Gentleman will acquire a Right to attend.

tend the Lectures and Dissections as long as he pleases; and such, besides being taught the Arts of Dissecting and Injecting, may, at the Beginning of the Summer, have an Opportunity of applying the Bandages upon a Machine, and of copying them if they think proper; and likewise of performing the Chirurgical Operations.

And as there are many Gentlemen who, although they do not belong to the Faculty, yet would wish to attend Anatomical Lectures occasionally, in order to acquire a Knowledge of the Animal OEconomy, such may be admitted every Year to the Lectures, at all Times, for Five Guineas.

Gentlemen who have already studied Anatomy, are settled in London, and do not chuse to go through a Course regularly, but only to attend the Lectures occasionally, may likewise become perpetual Hearers for Five Guineas.

Such

Such Gentlemen as have already subscribed to One or Two Courses of the Lectures given by Dr. *Hunter* and Mr. *Hewson* in Partnership, may be admitted on the same Terms as if they had subscribed to Mr. *Hewson* singly. And those who have already subscribed for Four Courses, given during the Partnership, shall be considered as having a perpetual Right to Mr. *Hewson*'s Lectures, and their Visits esteemed a particular Honour.

Two Courses will be given during the Winter Season, the One to begin about the First of *October*, and to be finished about the Middle of *January*; the Second to begin about the latter End of *January*, and to be finished before the Middle of *May*.

A Syllabus of the Course will be given to each Subscriber, that he may at all Times know what is the Subject of the Lecture.

www.ingramcontent.com/pod-product-compliance
Lightning Source LLC
Chambersburg PA
CBHW031750230426
43669CB00007B/562